CHILDREN IN NEED
OF SPECIAL CARE

HUMAN HORIZONS SERIES

CHILDREN

IN NEED OF SPECIAL CARE

THOMAS J. WEIHS

Souvenir Press

Printed in Great Britain by
J. W. Arrowsmith Ltd,
Winterstoke Road, Bristol

PREFACE

THIS book represents an attempt to describe childhood handicap not only as a pathological condition which has to be remedied, or better, prevented, not only as a need which has to be met, but equally as a challenge to the self-knowledge and development of the educator, therapist or parent. In this book, I should like to argue that help can accrue to the handicapped from a change in our understanding of them and in our way of being.

My approach is directed to persons in their encounter with a handicapped child. I do not want to persuade of proven or documented facts, but to stimulate personal, subjective experience in relation to handicapped children. I have, therefore, chosen not to support any of my descriptions or statements by references to specific literature in the text, but have compiled a bibliography at the end of the book.

My own work with handicapped children has been inspired by the teachings of Rudolf Steiner and by the late Dr. Karl König, the founder of several centres for handicapped children and adults in this country and abroad.

This book is the fruit of thirty years of intimate living with handicapped and disturbed children and of teaching and treating them. That it could be written is due to the untiring help of my wife and of my friend, Graham Calderwood.

THOMAS J. WEIHS

Spring, 1970
Camphill, Scotland

CONTENTS

Preface   5

I
INTRODUCTION AND GENERAL ORIENTATION   9

II
INTIMATIONS OF CHILD DEVELOPMENT   22

III
ASPECTS OF DEVELOPMENTAL HANDICAPS   39
  1. Morning and Evening   40
  2. Left and Right   47
  3. The Palsied Child   56
  4. The Restless Child   64
  5. Autism   74
  6. The Blind Child   100
  7. The Child with Impaired Hearing   109
  8. Children Suffering from Aphasia   115
  9. The Emotionally Disturbed and Maladjusted Child   123
  10. The Education of the Retarded Child   130
  11. The Mongol Child   136

IV
THE ENVIRONMENT OF THE HANDICAPPED
CHILD   141
  1. The Family   142
  2. School as Environment   149
  3. Adult Communities   168

Bibliography   176

I

Introduction and General Orientation

OUR CONCERN is for the child whom we love but of whom we despair. He may be my own child or a child entrusted to my teaching, or the child I encounter in my friend's house. I want to love this child. I want to see his good sides. Yet—why is he as he is? Why can he not be as other children are? Is it because he can't be, or because he doesn't want to be? Why must he always do things which are so annoying and hurtful? How can I love him? How can I cope with the anxiety, fear and frustration he causes for me? Are there other such children, or is he the only one? Is he one of the many, many thousands of handicapped children who attend special schools or training centres? Does he belong to any particular group? What is the diagnosis of his condition? Can he be cured? What is the right treatment? How can he be educated?

All these questions arise and immediately place the problem into the more general aspect of classification rather than into the purely individual one. Here, it is interesting to note that although the problem of classification is a universal one in the civilised world, it differs from country to country, not so much from the diagnostic viewpoint which is possibly more international, but very much from the viewpoint of administration.

In most countries, including Britain, a differentiation is made between physically and mentally handicapped children. A third qualitative classification consists in a group of so-called maladjusted children. There can, of course, be combination of the three classifications in any one child. In fact, it will become apparent that the distinction which these three classifications present are actually not applicable to the individual child but are abstractions referring to groups of handicapped children. We shall, however, try first of all to clarify the meaning of these three principles of classification in the light of our concern for the individual child.

The differentiation between physical and mental handicap has for a long time been supported by an old idea that "primary" mental defect is something in itself, independent of physical or organic background. This idea has fostered the attitude that the physically handicapped child is merely bodily inferior to the normal and healthy, while the mentally handicapped child, on the other hand, is inferior spiritually and as a person, any physical difference being an expression of this inferiority.

We no longer accept the existence of "primary" mental defect, just as we no longer accept the antique "Fever" as a medical diagnosis. Unfortunately, the spectre of the outdated concept of "primary" mental defect still haunts minds and many parents and their children are victims of it.

The other classification, that of "maladjustment", which means a disturbance in relationship to the environment, has fortified the belief that the *mentally* handicapped child is objectively handicapped and that the relationship between the child and his environment is at best, only of secondary importance. This attitude, however, minimises the role played by the environment in a child's problem. The same would hold good for the physically handicapped child and it will be demonstrated that this belief constitutes an obstacle for the full understanding of the problem of the mentally handicapped child.

But the individual child who has so deeply aroused our concern does not seem to fall neatly into any one of the three groups mentioned. He so frequently seems to be incapable of doing what his years demand. Is it because some degree of *physical* handicap lies at the root of his inability? His movements do appear to be less finely co-ordinated and skilful than might justifiably be expected. They are often odd, a little clumsy, and sometimes even bizarre. Yet, he is not paralysed, neither is he blind or deaf. There seems to be no obvious particular physical defect or handicap and still, in respect of his intelligence which might be supposed to be unaffected, he has certainly failed to learn to read or write. In spite of this, he often displays flashes of intelligence. He seems at times very alert and in fact, understands well what is expected of him, well enough to do the opposite. Nevertheless, he is not maladjusted in the sense of delinquency, although his motivations appear to differ from those of other children of the same age in the same situation. It may seem as though he were bent on irritating, provoking, even tormenting his surroundings, and the problems he can generate in his family and at school can hardly be described as being different from those of social maladjustment.

Clearly, then, the abrupt categorisations into mental handicap, physical handicap and maladjustment, useful as they are for administrative purposes in education, would seem on closer examination to be irrelevant to an individual handicapped child and his situation.

Further, a child who may have sustained a crippling infection of poliomyelitis in his early years and whose legs are so badly paralysed that he has to walk on crutches or that he is confined to a wheel-chair, will naturally strike us as being physically handicapped. His most pressing or central personal problem, however, may be essentially emotional, although he may not demonstrate it. We, in turn, may not be sufficiently sensitive to his problem, because we are overly impressed and distracted by a physical handicap.

Equally, a child who fails at school and who is so mentally retarded that he may not even be able to keep up with the education offered at a special school, may well have an organic handicap or deficiency at the root of his problem. He may have suffered from an early inflammatory process in the brain or from an organic condition that impairs the flow of functions in his central nervous system, but seems to be perfectly normal and fit physically. He is, perforce, classified as mentally handicapped only. Yet, a physical lesion of the central nervous system *has* occurred, and while it may not be as striking as a wheelchair case, in what remote sense can the idea of physical handicap be dispensed with as irrelevant in such an instance?

The third classification of maladjustment denotes a breakdown between the child and the social demands with which he is confronted. Now, this in fact may often be the outcome of some physical handicap, or of a peculiarly constituted personality, or merely the reaction of an individual to an unsuitable environment. A defect which may partake in all three of the administratively accepted categories cannot be in only one of them—that of maladjustment. Any such assertion confounds the facts and further highlights the rigidity of the official classifications.

Having touched on the *qualitative*, we can now turn to the *quantitative* aspect of classification and we find the idea of testable intelligence prominent.

Towards the end of the last century, the notion appeared that there could be a curve describing distribution of grades of intelligence as there is one illustrating the variations of the physical height of the human body. The term "mental test" was coined and scales of ability were carefully worked out against which school children were "mentally" measured.

It is not to be wondered at that in an increasingly mechanistically orientated society, the development of a scale for measuring mental ability was seized upon with enthusiasm, with the result that practically all over the civilised world, the concept of "mental age" was to become widely accepted.

This concept is again administratively useful, but unfortunately, "mental age" has been misinterpreted and misrepresented as the only age that matters, whereas it is only significant in respect of testable intelligence in the same sense that growth is significant for physical prowess, and just as no one will insist that height is a measure of physical chronological age, it does not seem reasonable to insist that mental age is a measure of levels of maturity or development.

It has been a matter of dispute whether a constant factor can be ascertained in testable intelligence or whether it is subject to change in the course of a child's development. This particular controversy has been brought about by trends in modern educational science. We must accept the validity of a basic constancy in intelligence quotients in so far as the scales for testing intelligence are so devised as to assess average intellectual development in the average child. These scales demonstrate the constancy of the development of intelligence or the constant factor in the chronological advancement in a child. It is a matter of course that this can only be ascertained as an average, and that the development of the individual will always vary from the norm.

It is likely that the constant factor in measurable intelligence is more obvious and more definite than the constant factor in the increase in physical height and weight in the average child. The important question is: *What* is actually measured by intelligence tests?

Constancy in the development of intelligence and its continued progression go on only up to the fifteenth or sixteenth year and are important points worthy of consideration. Can it be intelligence that we are measuring, if no increase is noted in it after the fifteenth or sixteenth year of life? No one would hesitate to deny that his intelligence stopped growing after this period, meaning intelligence in the ordinary usage of the word. It is, therefore, obvious that testable intelligence is something different and more specific than that which is usually meant by intelligence or reason. Moreover, reason, or more comprehensively, personality, are things usually associated with the idea of mental age. Here lies a crucial problem and it is necessary to attempt a realistic appreciation of what it really is that is measured by the ordinary intelligence test.

It is likely that the qualities measured are of a purely neurological nature and that the nearest we can approach to them in ordinary terms is to describe them as motor-qualities of the finest and most differentiated kind. We shall return to this in a later chapter.

The query now arises as to why the measuring of these particular qualities should have come to play such a dominant role in the assessment of handicapped children and their education. Be it said at the outset, the importance of intelligence testing is least significant as far as *handicapped* children are concerned.

In a scientifically orientated society, education is regarded as one of the sciences rather than as an art.

Education as an *art* cultivates and guides a child as a growing *person*. Here, testable intelligence is of little relevance, which is borne out by the fact that it becomes stationary in adolescence just when the maturing of personality increases at a rapid rate.

Education as *science* imparts information and develops skills in children of school age which is, no doubt, an important aspect of education, but by no means the only one.

Testable intelligence has been invested with excessive importance because this particular faculty of the human mind seems to fit better to the theory of information, and men so trained and orientated can adapt to computerised automation more readily. Hence, education for industry and public services becomes increasingly based on norms set by the assessment of intelligence.

As the idea of average distribution is as applicable to intelligence as it is to height or weight, it follows that the distribution curve shows that a certain proportion of the population has intelligence quotients on its upper or lower fringe. It follows that those parts of the population which constitute the lower fringe in the distribution curve of intelligence are made socially more handicapped than otherwise necessary. This is demonstrated by the fact that in times of severe unemployment, a proportion of the population becomes institutionalised in hospitals for the mentally sub-normal, whereas in times of full employment it is re-absorbed into the general employment market.

It is this that causes the apparent handicap in a child of lower than average intelligence and wrongly renders testable intelligence an all-important criterion.

The predominently social importance of testable intelligence is further demonstrated by the following interesting fact. Quantitative evaluation of educability differs in different countries. In Britain, we reckon I.Q.s from 80 upwards as normal for educational purposes, I.Q.s between 50 and 80 as educationally sub-normal, requiring special schooling, and I.Q.s below 50 as severely sub-normal, although trainable in the higher ranges of this category.

In the United States, the quantitative classification is different in that children with I.Q.s down to 68 are regarded as "slow learners" and are still included in the streaming of ordinary education. From I.Q.s of 68 downwards, there are four groups of retarded children: mildly retarded, medium retarded, severely retarded and very severely retarded. In American terminology, "retarded" corresponds to the British term "handicapped" or "mentally handicapped".

At this point, the following question must be asked. Can we really mean by a retarded or handicapped child a child who is simply less intelligent or duller than an average child? Are all the very many children with I.Q.s below 80 in British special schools handicapped children? Many of them have the same intelligence quotient as their parents and grandparents had. Many will become respectable citizens with their own families, taking responsibilities in their own social sphere, in work as well as possibly in public life. Some children may have I.Q.s as low as 60 or 50 and still learn to manage their adult lives in a human and satisfactory way without help.

In point of fact, has the assessment of handicap according to intelligence any reality at all? Are there perhaps *no* handicapped children? Or—are only the very stricken children with additional severe physical defects to be described as handicapped? It would seem that this is not a quantitative matter.

The Chief Medical Officer to the Ministry of Education made a most important distinction as much as 17 years ago. He stated that a differentiation would have to be made between children suffering from an innate defect of intelligence (in my own description, those normal and healthy children who are less intelligent owing to their genetic background or lack of social stimulation, but who achieve their own level of intelligence which happens to be less than the average), and others who suffer from an intellectual handicap on the basis of extrinsic causes (who would be truly handicapped children).

The Chief Medical Officer stipulated that the former group be educated according to their testable intelligence, whereas the latter group require "remedial education".

Strangely enough, this very fundamental piece of insight on the part of the Chief Medical Officer at the time has not been put into effect to any adequate extent. In spite of serious attempts to find other, more human, personal and individual means of assessment, in spite of doubts that have arisen in scientific circles as to the constancy of testable intelligence, and in spite of a general inclination not to want to over-estimate the value of intelligence tests, testable intelligence has remained the basic criterion in administrative classification of mentally handi-capped children. Although there is awareness that the three categories—mental handicap, physical handicap and maladjustment—intermingle and intersect, they are artificially maintained as fundamental principles.

But what alternative, more pertinent and more valid classification could possibly be found? In approach to this problem, I should now like to bring up the idea of *development*, and to suggest that an attempt be made to find a way to assess a child's condition, not on the basis of his faculties, of his measurable performances, skills or abilities, but on the basis of his development.

There is a crucial difference between those children who have an innately lower-than-average intelligence according to the law of distribution, and those children who originally had a higher potential, but suffered from some early general or possibly more specific set-back, so that their testable intelligence is noticeably lowered.

In schools providing remedial education, we always find a con-siderable proportion of children with relatively normal or even good intelligence. These are often children who, after an inflammatory condition affecting the central nervous system, are left with an intelli-gence quotient of between 90 and 110 that put them into the average section of the population, but who, had they not suffered from such an illness, would have scored considerably above the average. These children are often severely handicapped, in spite of possessing an average I.Q. because it does not relate to their own innate intelligence but has been imposed upon them by a developmental disturbance, and they often fail to make their way, even in special schools.

If we suppose that the assessment and classification of handicapped children should be based on developmental aspects, testable intelligence becomes only one of the many physical or mental factors to be taken into consideration. The development of motivation, social adjustment and emotional life become all of equal importance to testable intelligence or "mental age" in the maturation of a personality.

We may propose that the study of developmental failure would have to become the basis of the assessment of handicapped children, and to this end, we shall require an understanding of child development from the point of view of pathology.

As an illustration, we shall now consider four diagnostic syndromes in child handicap, namely, cerebral palsy, post-encephalitis, mongolism and childhood autism.

Cerebral palsy has played a great and important part in the changing attitude to handicapped children. This is probably largely due to the fact that a fairly severely palsied boy in the United States became a doctor of medicine between the two Great Wars, and took up the cause of the palsied child himself. Dr. Phelps made people increasingly aware that many cerebral palsied children, even some of the most incapacitated, were highly intelligent and that if one learned to communicate with them, one could participate in their, often superior, sensitivity of mind. Here, because of the physical handicap, the intelligence test plays a helpful part, as it allows an objective scientific demonstration of the intellectual potential of these children, despite the fact that they can make no overt, meaningful use of it.

The stir caused by this new recognition of the mental integrity of many cerebral palsied children resulted in the inclination of physicians all over the world to diagnose all handicapped children as cerebral palsied. It was the one form in which childhood handicap became acceptable and aroused the compassionate interest of the public, allowing for the inclusion of handicapped children in educational and social activities. It is likely that hundreds and thousands of handicapped children today owe a debt of gratitude to the palsied child, and certainly to Dr. Phelps. Their lot would otherwise have not become the compassionate concern of the public to the same degree and at the same rate as it has today.

Let us now consider the medical-diagnostic aspect of cerebral palsy. No one doubts that cerebral palsy is due to a particular type of brain damage, usually sustained at birth through some form of birth injury or anoxaemia, which is an oxygen starvation in the brain. Such damage can also occur pre-natally.

Although these aspects of the causation of cerebral palsy seem unequivocally established, a puzzling and striking phenomenon manifests in these cases. The cerebral palsied child by whom brain damage has been sustained either before or immediately following his birth, nevertheless shows no noticeable signs of palsy during the first weeks or even months of life. It requires considerable expertise to be capable of diagnosing cerebral palsy within this early period.

How is this to be understood? Why does an injury to the brain, occurring before or at birth, not manifest until so much later?

In fact, it is really only in the course of the first year of a child's development that palsy becomes overtly manifest. Many palsied children who fail to learn to swallow their food properly or to drink from either cup or straw are reported to have sucked normally as tiny infants. Equally, children with the severest spasticity who do not learn to walk or to move any limb in a co-ordinated way, all appear to have kicked and moved normally during their first weeks.

There is another strange phenomenon not often noted by people who have little intimate knowledge of cerebral palsied children. It is that while the most frequent form of cerebral palsy, that of spasticity, seems to grip a child throughout the day, it leaves him the moment he goes to sleep. Sleeping spastic children appear relaxed with no sign of contraction or deformity. (Deformities that do not relax during sleep can develop later as a result of contractions, as in the case of an arm or leg which has been in plaster too long and may become fixed in that particular position.)

These striking phenomena are largely so puzzling because of some misinterpretation of the term "cerebral palsy". The word means "brain paralysis", which implies an inability to move due to or in connection with a disfunction of or injury to the brain. However, movement itself is *not* imparted by the brain. It rather imparts the control, guidance or direction of movement. In distinction from the non-cerebral, peripheral palsies, there is no real lack of movement, no muscular inability in cerebral palsied children, but rather an inability to control muscular contractions. Once we realise that cerebral palsy is a failure of the cerebral control of movement, we begin to understand why palsy becomes recognisable only some time after birth, and why in sleep, the defect disappears.

A child is born with a considerable amount of involuntary movements, amongst them the so-called kicking movements and a number of basic reflexes. The earliest and most established reflex is that of sucking, but there is another interesting reflex which can be seen in young infants. When an infant is held up under the armpits and allowed to touch a solid surface, such as a table-top with the soles of his feet, he can be made to pace after a fashion. He will set one foot in front of the other, but always with a tendency to cross the legs, so that a kind of scissor-gait results. In normally developing children this early infantile reflex has ceased by the time they begin to stand and to walk. The failure to overcome this reflex and to establish voluntary motor-control in the early months *is* a typical form of cerebral palsy.

We shall turn now to post-encephalitis, which presents a particularly telling example of the difference between medical and developmental insight. The term "post-encephalitis" designates a pathological condition which is a consequence of encephalitis, an inflammatory process in the brain, usually due to some form of infection or possibly to an allergic reaction to vaccination or immunisation.

Encephalitis in infancy usually results in the hyperkinetic syndrome, which is associated with extreme restlessness, possibly destructiveness, self-aggression, over-breathing and severe mental retardation. In a young school-age child, the post-encephalitic picture is rather that of moral disturbance. Here, intellectual development is only slightly impaired, if at all, whereas ethical difficulties like lying and stealing are so predominant that the condition used to be referred to as "moral insanity".

Post-encephalitic adolescents and young adults usually have the symptoms of loss of memory and lapses into infantile behaviour patterns, while in the mature adult, the post-encephalitic syndrome is that of Parkinsonism, a slowing-down of all movements, a mask-like face, with finger-tremor and a tripping gait.

While encephalitis as an infectious disease has definite and recognisable symptoms, and the course the disease takes depends on the severity and localisation of the infection, on the constitution and resistance of the patient and on the available treatment, the *post-*encephalitic syndrome is not so circumscribed but manifests in widely varying forms of handicap, showing little resemblance to one another, and specific only to certain age-groups and different developmental phases.

In childhood, relatively small differences in the age of the onset of the disease cause separate forms of handicap. Once adulthood is reached, even great age differences hardly influence the ensuing handicap. This demonstrates the fact that childhood handicap, while often *caused* by an illness or impairment of organic function, is *not* the illness itself, but rather a consequence and a deviation or derangement in a child's further development.

Now regarding mongolism, which is probably the largest single diagnostic group among childhood handicaps, there is something that has been a puzzling phenomenon for a long time and many things about it remain unclarified to this day. During the last decade, however, it has been demonstrated that mongolism is linked to some genetic pathology, and a good deal of research has been carried out as to the mechanism and causation of mongolism.

It has been discovered that mongol children do not only have specific physiognomic features, but that each single cell in their bodies shows a distinctive structural character. Although the mechanism of this structural difference is known, its cause has not yet been fully understood, and only in rare cases, is this genetic aberration of a recessive hereditary nature.

Intriguing as the genetic aspect of mongolism certainly is and notwithstanding its obvious importance to scientific understanding of the phenomenon, it again does not help us in our encounter with an individual mongol child.

There is, however, something which may help us in an endeavour to relate ourselves to the mongol child with understanding. It is this. By and large, mongol children have physiognomical and morphological features that resemble the human embryo around its second month. This will be described in detail later in the chapter on Mongolism.

At this point the question may be asked whether or not mongolism is in some way a halting of differentiation in development during the early embryonic period. Here again, we encounter the difference between the medical and the developmental approaches.

The fourth and last diagnostic class to receive consideration here is infantile autism, or childhood psychosis, as the condition was originally called in Britain.

The classical symptom of childhood autism is evasion of encounter with another person, often to the extent of visual and auditory avoidance, while interest in inanimate objects persists and can be highly developed. Speech may be absent or manneristic, often limited to echolalia or showing transposition of personal pronouns (saying "I" for "you" and "you" for "I"). Anxiety, dependence on sameness, bizarre movement-patterns and obsessional behaviour are often found in psychotic or autistic children, and their intelligence tests show a particularly wide scatter of points between failure and achievement.

While the other three diagnostic classes—cerebral palsy, post-encephalitis and mongolism—have aspects of causation or aetiology that are generally agreed upon, this is not the case in childhood autism. There are two common and apparently contradictory approaches to the aetiology of this condition; in one approach, it is regarded as being psychological in origin, being a mode of reactive behaviour to unfortunate or pathogenic environmental circumstances, while the other school of thought sees autism as the result of a metabolic aberration or as a form of brain pathology.

There are, no doubt, reasons for considering the psychological as well as the organic aetiology in the syndrome of childhood autism, but we hope to be able to show how the apparently irreconcilable contradiction between the two may to some extent be overcome.

In our approach to the autistic child, we shall try to demonstrate his problem as a failure in the development of motivation and in the experience of Self appropriate to his age, as well as a breakdown in the experiencing of another person as an individual possessing an ego.

To recapitulate, the diagnostic classes dealt with up to this point would ordinarily be placed in the social classifications of physical handicap, mental handicap and maladjustment, and yet it remains questionable as to how meaningful the distinction of physical handicap, for instance, would be for a child so completely incapacitated by cerebral palsy that he can neither move, nor develop speech, although he may understand speech. Such a child is so severely handicapped that he may suffer severe emotional unbalance, which may generate problems far beyond those of so-called pure mental handicap.

Likewise a mongol child with a relatively low I.Q. but good adaptability may be able to function on a very much higher level than the severely physically handicapped child, and the autistic child, who may be classified as maladjusted and who may have a fairly *high* testable intelligence, may be so severely disturbed in his motivation that his ability to function in the human world is far more reduced than that of a mongol child with a *low* I.Q. and so forth.

I hope I have shown that the usual differentiations of medical diagnosis give as little real guidance in forming an attitude to an individual handicapped child as do the socio-administrative classifications. Of course educational and social services must be based on administrative classifications of handicapped children in the population. Equally, medical help has to build on thorough investigation of causes and aetiology. Nevertheless, I shall leave these aspects alone after my cursory mention of them, as they seem to be incidental to our own task, which is the attempt to learn how to understand a handicapped child.

I shall now try to pave the way for this understanding on a developmental basis and the next chapter will describe images of child development which are not intended to be scientific, but which are intended as a help to call forth a compassionate personal attitude, while we learn to live with and to love a handicapped child.

II

Intimations of Child Development

WE ARE accustomed to saying that children "grow up". But is this a truly accurate expression? Does a child really grow *up*?

When we try to see a little new-born child as it really is and not as we imagine it to be, we see it perhaps in the arms of its mother or in its cradle or cot. We never find it on the ground or on the floor or on the earth.

In contrast, the new-born calf or foal is dropped, and falls at once to the ground at birth. It is a lump to begin with, wrapped in glistening slime. It is licked dry by the mother animal, after she has eaten the sheaths and afterbirth.

The new-born animal soon begins to wriggle and struggle onto its legs, which are long and relatively sturdy, whereas the trunk and head in both a calf and a foal are slight in comparison to them.

It is delightful to watch a new-born calf in a meadow or in the byre struggle on to its limbs, sway a little on its four sturdy legs, stumble, then straighten up and finally, stretch its neck to reach the mother's teat, in order to suck.

Calves or foals literally grow *up*. From the first moments after birth, movement begins in the periphery, in the limbs, then spreads into the trunk, raising it from the ground. Finally, the head itself is raised, so that the new-born animal stands firmly on its four legs, well-ensconced, and reaches up to the mother's teat to drink. With every hour and every day that passes, the young animal grows up and raises itself ever further from the ground.

But is this the case with the human child?

Have we ever observed a new-born human child wriggling up from the ground, getting onto its legs, raising its head and sucking?

Do we not, to begin with, see the human child lifted up in its mother's arms or placed in its cradle? Does it not take a fairly long time for the child to begin to reach the ground, the floor? Do not six, nine, perhaps even twelve months pass before the child "comes down"? Would it not be more correct altogether to say that a child grows *down*?

This is the first question that arises when we observe children with a sympathetic and open mind, as though we were seeing them for the first time.

A child is born amidst the trumpet blasts of intense labour and pain. It is thrust out of the dark, enveloping womb into the cold, open spaces of this world. It draws its first breath, cries its first cry, and after the intense effort of being born, the infant sleeps.

When it wakes, its hands wriggle, its legs and feet kick, its eyes roll independently of each other. At times, the infant will cry, mostly it will sleep, waking only to drink.

The new-born infant is helpless and totally dependent on its mother. The single co-ordinated movement it has at its disposal is the sucking reflex. Yet, within its first days and weeks, control begins to descend over the sea of movement in which the new-born child seems to be adrift.

We shall now do something that can only be done in imagination. We shall observe the development of the child's movement-patterns over an entire first year. We shall disregard all other experiences garnered from the encounter with a very young child and try only to concentrate on what may be observed in the unfolding development of motor-control, of the voluntary co-ordinated movements which a child brings to completion in the course of its first twelve months.

We said that the child was, as it were, immersed in a sea of uncontrolled, spontaneous, unco-ordinated movements. (Anyone who has seen palsied children can recognise elements of early infantile movement-patterns in some of the strange, writhing movements of athetotic palsies.)

The very first motor-control we observe in a child, frequently only a few days after his birth and not later than a few weeks after, is co-ordination of the eyes. An infant learns to co-ordinate the movement of both eyes so that he can focus on one particular thing. He learns fairly quickly to follow a moving object with his eyes. In the course of his second month, his head and face begin to follow the direction of his gaze. An infant learns to turn his head in the direction of a light or sound which he experiences, and generally in the course of the first three months, an infant gains distinct control over his movements of head and neck, so that in a period when the rest of his body is still steeped in an ocean of involuntary, spontaneous, kicking and wriggling movements, the head emerges, having achieved its own control and established the loftiness and dignity peculiar to the head.

In the course of the next few months, the voluntary motor-control of a child descends from the neck into the shoulders and arms. The period up to the sixth month is characterised by his learning to grip with his little hands those things he has so far only "gripped" with his eyes. He can hold objects, handle them, move them gradually from one hand to the other. His two hands can clasp each other, and while he achieves voluntary control over the movements of his arms and hands, his motor-control also descends from the shoulders down over the trunk to the hips.

Then the child sits up, although only with support to begin with. This mastered, he is in the position to have a certain limited, personal space around him which he can control with the combination of gaze and grip of eye and hand.

Usually at the very beginning of the second half of his first year, a child has acquired the ability to sit freely and to master his own little kingdom of motor-control as far as his arms and hands can reach. In the course of the next few months, voluntary co-ordination descends still further into the thighs, and by the seventh, eighth, or ninth month, a child has learned to raise himself into a crawling attitude, then to crawl, to move along with the help of arms and legs, or to shuffle along on his bottom, using co-ordinated movements of arms and hands as well as of legs to propel himself. Usually by the ninth or tenth month, a child will, for the first time, pull himself up into the upright position.

Thus it is seen how motor-control descends from the knees of the infant into the lower legs, ankles and feet and how finally, towards the end of the first year, the upright position is achieved. A child now stands freely on his own two little feet and attempts his first steps.

Watching a little child take his first steps is a moving and impressive experience. We should only compare it to our own getting up out of a chair and walking, or getting up out of bed after a long illness to walk again for the first time. We carefully put down our feet, lever ourselves to the edge of the chair or bed and after stretching our feet forward, we push our weight on to our legs from behind. How very different are the first steps of a little child! He stands freely, his relatively large head moves forward first, his neck, shoulders and trunk follow. His legs and feet remain rooted on the same spot. Just before the child falls forward, he advances a foot and with a tip-toe, he launches into the first two or three steps before sitting down again.

What have we actually witnessed in the events that culminate at this point? We have witnessed motor-control descending from the infant's eyes through his whole body to his feet. We have witnessed the motor-descent of a young child. A child has *grown down*.

In the preceding description, mention has been omitted of two of the most highly differentiated and controlled motor-developments of the first year. The first is a baby's smile, which appears from about the sixth week onward, when his head begins to free and establish itself, and which is one of the most moving phenomena of early child development. We shall return to this fundamental achievement later on when we look into the development of *experience* in a young child. The second is a development of controlled movement manifesting itself in babbling, in "baby talk", which appears about the same time as when a baby has learnt to grip. Babbling continues throughout the second half of the first year, and an extraordinary amount of effort, practice and intensity is put into this by a child of that age.

Babbling is often mistaken for the beginning of speech. As babbling is composed of syllables which in certain sequences would seem to form the first words, the most typical of which are me-me-ma-, da-da-da, this is understandable. From another point of view, it is fairly obvious that babbling is another facet of motor-control which, in a sense, starts at birth with the first cry and in the second half of the first year, enters into a phase of intense practice of the motor-speech organisation, which shows itself in the form of babbling and the formation of syllables.

One of the most striking and impressive discoveries in the field of child development of this century is the fact that baby talk does not vary according to the language of origin of a child, but is common to all babies. There is no baby talk in English, French, German or Russian, not even in African, Chinese and Eskimo. Hence an archetypal form of "language" would seem to unite all humanity in early infancy.

A child loses this "universal language" when, in connection with his achievement of uprightness in the course of his second year, he begins to acquire his mother-tongue.

The Book of Genesis indicates that mankind had an early universal language. This was lost in the attempt to erect the Tower of Babel, a structure of unlimited size and complete uprightness, which was to reach from earth to heaven, and which resulted in the fragmentation of the original universal language into many different tongues.

As we have broadly set a child's motor-development in the course of his first year of life, we shall now set out to observe his speech-development in the course of his second year.

A child's acquisition of his mother-tongue is, of course, not the result of a separate process from that which brings about his original baby language, but to begin with, we can proceed as if this were the case, and the way in which a young child begins to learn the use of his mother-tongue in the course of his second year shall now be considered.

Here, we can make an interesting observation. During the first months of the second year, a child uses nouns only, and then from about the fifteenth month onward a child begins to add adjectives to his nouns. This is the phase in which a child says things like "good Mummy", "nice doggy" and so on, and his first two-word sentences emerge as a combination of the nouns he has already been using with newly acquired adjectives.

Usually only by the end of the second year, from the twentieth to twenty-third month does a child begin to use verbs, and has thereby mastered the three basic types of words that constitute language. Nouns, adjectives and verbs are brought under his command in that order.

Here we see a second form of descent, not down through his physical organism this time, but down through the threefold organism of language, going from the *cognitive* quality of nouns, through the *emotional* quality of adjectives to the *active* quality of verbs.

In the course of a child's third year, his ability to think develops and unfolds. Through incessantly repeated percepts the new faculty of memory is generated in a child. Simultaneously another faculty develops. Imagination comes into play with memory, and from the interchange between these two faculties, questions crystallise, like forms swimming from an inchoate dawn. "Why?" "When?" "Who?" "Where?" Such questions as these lead up to a child's experience of his own individuality, to the dawn of his own Ego.

To sum up what has so far been set forth. It has been seen that the descent of motor-control during the first year of life occurs from the head downwards over the trunk into the limbs. In the second year, there is the descent of language-development from the cognitive through the emotional to the active, and in the course of the third year, there is a more subtle descent of the ability to think via memory, through imagination to enquiry, which culminates in the experience of the own ego.

Up to this point, we have regarded child-development from a one-sided point of view. We might say that we have spoken only of more overt degrees of motor-control or, in other words, of a child's *behaviour*.

We would now have to find out if we can observe the development of a child's *experience*.

Can we arrive at an appreciation of experience in early infancy?

To do so, we must return to what we called the descent of motor-development in the first year. It is obvious that as motor-control descends from a child's head to his limbs and culminates in free locomotion, a child's scope of movement increases considerably in the course of his first year, a trend which continues to increase in his second year, at which time he begins to romp and play and run about. At the same time, not only the ramification of movement is extended, but motor-control becomes refined and differentiated as a child gains mastery over the finer muscles of lips, tongue, larynx and mouth, all of which constitute his speech-organisation, and learns to speak.

Finally, in his third year, a child's thinking develops as a still higher refinement of movement in which we no longer speak of muscles, but of movement purely from percept, memory and imagination to concept, query and idea.

Thus the broadening of the field of movement is accompanied by its increased refinement. But this very refinement in turn constitutes a deepening of dimension, because the spoken word in the spatial sense can reach further than the action of hand or foot, and the developing ability to think not only embraces *space* but *time*, too, and reaches out into the past as well as into the future.

Although in the very young infant it may be difficult to observe anything *other* than its overt functions, by the time it begins to develop language and speech, we can hardly continue to isolate the development of movement from that of experience.

For in fact, speech is by no means merely the *ability* to speak; it is comprised of hearing, of the understanding of language and of more than the mere acquisition of words.

In order to appreciate how a young child proceeds from baby-language to that language with which he later communicates, we shall have to include the development of the senses on which his experience and ultimately, I submit, his consciousness are based.

Here, we are limited to a child's overt modes of expression, and therefore, before he has developed language which is understandable and interpretable by us, our estimation of the nature of his experience is bound to be beset by difficulties. But we have already encountered one manifestation of infant experience. That is its smile in response to our smile. This infant smile has puzzled observers considerably. It has been interpreted as the earliest form of imitation in the assumption that an infant *sees* the smile of the other person and imitates it.

Yet we see the first smile at a time when the infant does not recognise his mother in the sense in which we recognise someone or something we have seen before. Is the smile the expression of pleasure which it appears to be? and if so, what gives rise to the pleasure?

Just as little as an infant can account for the experience that causes it to smile, can we, as a rule, remember our own infantile experiences. There are, however, poets like Traherne, Jean Paul, Wordsworth and Adalbert Stifter who have given descriptions of very early infantile experience, and all these descriptions have one thing in common. They describe things not seen centrally, but, as it were, in a peripheral, omnipresent sense.

Helpful descriptions of early infantile experience also come from persons who, either under hypnosis or analysis, report that when as young infants they perceived their mother's smiling face above them, it was as though they were seeing themselves.

A number of years ago, some young friends of mine had their first child. I was taken into the room where the cradle was standing in order to admire the new arrival. I parted the curtains of the cradle and the infant—it must have been six or eight weeks old—began to smile at me with a radiant and intense expression of happiness. As I was not smiling myself, I was astonished, knowing that infants usually only smile in return to a smile. I was puzzled by the expression of delight, pleasure, pride and satisfaction that flooded over this infant's face. I did, however, sense someone behind me, and turning round, looked into the face of the radiantly proud father whose expression I had seen reflected in the infant's face.

Observations such as these, of which the foregoing is an example, can lead one to assume tentatively that the nature of consciousness in an infant and very young child is expanded and peripheral. This would mean that an infant does not experience himself as being *in* himself as the adult person does, but that he is one-with-his-mother and others in his intimate surroundings and therefore, smiles *our* smile and expresses *our* pride and pleasure.

Hence, it seems likely that infant consciousness in all its vagueness is universal rather than individualised, wide rather than narrow.

Should this be the case, new light is thrown on the development of baby-talk which, as I pointed out earlier is a universal language that transcends nations and races. The mother who knows whether her baby's cry is a cry of pain, hunger and frustration or satisfaction and well-being, will also know what African or Indian babies express.

Just as the infant gradually separates his own existence from that of his surroundings, he also gradually changes over from his global baby-language to his mother-tongue. By the end of his first year, together with his achievement of uprightness and walking, he can discern the characteristic tone-qualities of the language spoken in his near surroundings and can introduce them into his own evolving use of language. Thus he finds his way from an expanded periphery in which baby-talk originates to the differentiated and confined sphere of his own mother-tongue.

If a very young child's consciousness is more expanded than centred, if he is more generally of "mankind" than an individual, it follows that his learning is not an addition of new knowledge or skills, but rather the result of a protracted process of estrangement from a state of all-embracing consciousness.

In the beginning of this chapter, I argued that a child does not "grow up", but that his motor-control "grows down" through his body, then through language and thinking, while in connection with his sensory development, he is led "out" into a widening experience of space and time. We are now adding a third direction, assuming that a child's consciousness goes the opposite way from a condition of expansion to one of restriction and centralisation.

We shall now try to follow up the development of a child with reference to the dawn of his own ego-experience. A new-born infant appears to us as completely helpless and dependent, but we realise that he is dependent on an environment which, far from being separated from himself in his own experience, he finds included in himself. Albeit only in a dim state of consciousness, he *is* that very environment upon which he depends. Thus his first faint experience is not one of being helplessly dependent, but rather of being omnipotent. He himself is the one who feeds and stills his pangs of hunger and alleviates his own discomforts. This may be a strange and unusual interpretation of early infantile experience, yet it tallies with intuitive human experience as we have already said.

It is the withdrawal from an early state of expanded consciousness that gives rise to an experience of helplessness and dependence in a growing child during the first year of his life.

It is likely that in the course of these initial infantile experiences the foundation is laid for basic emotional or spiritual powers such as trust, confidence and faith. In this essential phase, a child becomes orientated in such a way that he will experience himself at a much later date as either isolated and meaningless, or as an integrated part of a meaningful and greater whole.

From this it can be concluded that religious feelings, ideas and convictions do not arise primarily from a longing to explain something or remedy an unsatisfactory situation, but rather from a dim memory of having once been at one with the entire realm of one's experience, of not having been in a world either hostile or friendly, but of having been oneself The Very All.

The gradual separation from the All which we originally *are* is of fundamental importance to the constitution of our ego. In the process of separation, the mother-infant relationship is naturally the predominant one. Breast-feeding and the feeding situation as such, bathing, toiletry, dressing and above all, holding and handling are the basic encounters in which this gradual separation from the original Oneness occurs. In these situations, the first ego-integration begins to dawn and it is crucial whether an infant feels held—held and *beheld*— by what he gradually has to learn to recognise and to accept as that which is not the "I" of himself, but ultimately the "You" of the Other.

In our time, it happens increasingly that young and immature parents become uncontrollably frustrated by the behaviour of their infants and this can lead to the tragedy of a battered baby. It is probable that behind this frustration lies the fact that such parents are confronted by the archetypal omnipotence of their infant and are unable to deal with it, because they themselves have lost their basic trust as a result of their own upbringing. They feel challenged and unbearably attacked by something in the infant which is sovereign, but to them utterly incomprehensible.

Ego-integration as a process, however, by no means reaches completion in the first year of a child's life. It continues throughout childhood and even into adolescence and adulthood. An infant's experience of his mother coming to comfort him is still one of his state of omnipotence. But the gradual development of mobility increases both his sense of power and independence, and the feeling of being separated.

While a very young infant's experience of his own little hand moving in front of his eyes against the light of a window or the sky is similar to our experience when we see the clouds passing over the sky, in the second year of life a child begins to see his own hand as a hand, *his* hand, perhaps already having perceived his mother's hand as a hand.

It gradually dawns on him that not only can he see but also that he can *be seen*. With this, the fundamental experience distils that there are *parts* of his own body he himself cannot see and that can only be seen by others. "Front" and "back" become new dimensions, by no means only in a spatial and objective sense, but as an existential experience of one's own way-of-being. Qualitative differentiations appear that have magic.

At the time when a child begins to unfold language, we can observe how his words have far wider and all-encompassing meanings, to begin with, which gradually diminish in scope until, in the course of time, they link up with ordinary adult usage.

A child does not initially learn *adult* language, but seems to go through different language-forms that may well have similarities to earlier forms in mankind's evolution. More clearly than in a child's language-development can it be seen in his play how factual objective elements begin to be integrated into his personal world of fantasy and magic. A piece of wood can be a boat, a house or father. Anything can be anything. Any kind of material can be animated in a child's play. A child is the Creator. He is the parent of his children, a doctor who heals patients, a builder of cities. But at the same time, he becomes increasingly aware of his dependence on his own makers who are his parents. A whole new world of triumph and conflict opens up in the growing child.

To illustrate this, I know of someone who as a little girl was "creating" a world of mountains and rivers in the sand on a beach in the fast fading light at the end of the day. She remembers just having had the primary experience of herself as identical with the Omnipotent Presence who created the world, and laid her hand upon the mountains she had created. At the same time, she felt anxiety lest she should not find the way up to the hotel in the dimming light, and with her other hand, reached out for the hand of her nurse who was nearby.

Child-development is a dramatic journey that proceeds through archetypal stages from a universal world of meaning, magic and *mythos* into an adult world of facts and causes.

Early child-development would seem to bear a relationship to the history of mankind as experienced by ancient peoples and expressed in their mythologies and religions. For example, the Fall of Man as described in Genesis (chapter ii. 7–10) is an archetype of that drama which takes place in all children when they know that they not only see but *are seen*, just as immediately after Adam's eyes are opened and he sees, he knows that he *is seen* and is naked. As another example, we have already mentioned the Tower of Babel and the confounding of language that accompanied it as an archetype for the differentiation of the various languages out of an original global baby-language.

I should now like to gather up what has been presented in the form of a simple polarity.

As against the development of a child's abilities, performance and behaviour, we saw an opposing movement in the development of a child's experience or consciousness, which goes from an expanded but vague and dim state to an ever-narrowing, evermore centred, differentiated and focused one. How can these two movements be reconciled?

It could be said that at birth, only the body of an infant is born. Once the navel cord is cut, an infant arrives as the discrete culmination of a time-extended genetic background. But this is true only of his body. With regard to the *functions* of that very same body, he is by no means as yet "born". His locomotion is still entirely dependent on his mother. In fact, his entire potential is still largely *unborn*. This not-yet-being-born can possibly be seen in an infant's faculty of experience, which is still within the womb of mankind.

It would be important to see birth as the severance or separation of an individual from a greater origin.

Two trends have been encountered in child-development—that of a physical birth, which leads to functional expansion, and that of the development of consciousness which leads from expansion to contraction or concentration. These two trends are characteristic of the way in which we experience ourselves, namely, our physical-organic constitution as against ourselves as a person.

Although these two extremes make up the totality of our human existence, they seem to appear anew in the light of the following picture.

Child-development can be seen as the integration of two processes, one stemming from the hereditary, genetic background and starting with the separation of the individual body at birth. The other is connected to the mythological-historical evolution of the human race, which does not start from a centred, individual, isolated beginning but from far peripheral reaches.

Progressive integration of these two processes takes place in an environment which is itself both physical and cultural, represented by mother, father and family to begin with, and later by teachers and others in addition to the institutions, conventions, and traditions into which the child is placed by birth, as well as the geographic, climatic, and biological conditions that obtain.

If it can be acknowledged that there are these three factors in child-development, the question arises as to what way and to what extent they contribute to childhood handicap. What possibilities do they hold for the understanding of and help to the handicapped child?

I should like to put forward the following views, less as an intellectual argument than as a moral commitment. The genetic element in child-development is likely to be seen in the variations in height, physical proportion, features and probably testable intelligence in the population. It seems, therefore, more related to those "normal" variations from the average as distinguishable from childhood handicap, which I drew attention to in the first chapter of this book. Regarding the extent to which genetics play a part in actual handicap, as in mongolism, the results of recent research seem to indicate that changes or pathologies in genetic structure appear in connection with environmental influences such as the age or state of health in the mother or other as yet unknown factors.

There is, however, another aspect where environmental and genetic factors interact. It is assumed that vicissitudes of physical environment exert a selective influence on genetic qualities through the "survival of the fittest". We must now accept the fact that the considerable improvements in hygienic conditions through technical advance counteract this mechanism, and thus cause an increase in the survival of genetic aberrations. This in turn gives rise to further technical efforts to influence chosen genetic survival, both by means of early suppression of aberrations as well as by means of selective biology. Such trends sow the seeds of distrust and fear with respect to science in its one-sided pursuit of causes.

I should like to suggest that the increase in handicapped children brought about by better chances of survival has to be appreciated and made culturally positive by a new understanding of the meaning of child handicap.

The physical aspect of environment seems to me to play by far the greatest part in the causation of developmental impairment. Although the advances in medicine and other techniques continuously eliminate pathogenic environmental factors, equally they create new ones, both accidentally as well as by wresting handicap from situations where death would previously have intervened.

The cultural and emotional aspect of environment is in danger of being driven into a comparative dilemma by being regarded too exclusively as a possible cause of child handicap. As a result, parents are too often forced into an attitude of uncertainty and feelings of guilt, while the positive potential of the emotional environment remains relatively unrealised.

An important and so far untapped potential of understanding and help can be opened up when the determination to see childhood handicap purely as an ill that has to be remedied can be relaxed. Only when what amounts to one of our preconceived ideas is set aside is a genuinely new interpretation of childhood handicap to be found.

An endeavour along these lines will be supported by the realisation that the third factor in child-development, which is the individualisation of consciousness, is of a different nature from the other two and one that is not so irreparably exposed and vulnerable to environmental vicissitudes. This third factor altogether would seem to retain a quality of unassailability and integrity, even while its manifestation is entirely dependent on the integration of the other two factors within the given environment.

The following analogy might better convey my argument. We might see the handicapped child as an artist who has to play on a faulty instrument. Even when we ask the finest of pianists to play on an unsound instrument that has been damaged or is out of tune, his performance will be a poor one, however gifted and skilled he himself may be. Those in his audience who have had little musical experience will think that he is a bad artist. Displeasure and disappointment will show on their faces and the artist will be frustrated and unable to give his best, but may, nevertheless, give a sham performance rather than expose himself.

On the other hand, if some people in the audience are musically sensitive and know the piece the artist is playing, they will perceive his intentions and interpretations. They may even derive intense enjoyment and satisfaction from his performance, because they sense its meaning and share the artist's experiences of the particular piece of music. The artist, noticing the facial expressions of his listeners, perceiving their indications of appreciation and understanding, will naturally be helped to give of his very best. His performance will improve and the "concert" may turn out to be a considerable and marked success.

The relationship between ourselves and a handicapped or disturbed child is of the same order as that between artist and audience. Inasmuch as we are interested only in the "instrument" and have no ability to perceive his personality or spiritual nature separated from the way his physical limitations allow him to express them, we present to a handicapped child an "unmusical" audience. Our failure to understand and appreciate him will of necessity not only affect his intentions to express himself, but also curtail his means of expression.

But if we train ourselves to regard that which is un-hurt and perfect in a child as that child's spiritual qualities, this in itself will foster his development and his ability to express himself.

Therefore, if we want to understand and to help a handicapped or disturbed child, on the one hand, we must realise that we are part of the environment in which this child has to live and grow, and on the other hand, we must try to see his behaviour, performance, abilities and inabilities in relation to what is always perfect in him—his experience of his own personality, his own ego.

In particular, we must not fall into the error of separating out his overt physical handicap and regarding it by itself. It must be interpreted in close connection with the process of mental and spiritual integration in him. If we succeed in this, we shall gradually learn to experience different forms of developmental handicap or disturbance in children as exaggerations of what we understand to be normal variations in human characteristics.

If we learn to find in ourselves the frailties and problems of a handicapped child, a new understanding and appreciation will open up for us, through which we can learn to love the handicapped child, not in an emotional sense or one of sentiment only, but as a means of extending help and support, and wherever possible, healing.

It is perhaps inappropriate to close this chapter without drawing attention to yet another aspect of development which is that of motivation and the acquisition of moral and social values by a child.

It has always been taken for granted that man's intelligence is socially orientated and integrated, since Aristotle's definition of man as a "zoon politikon"—a creature with political orientation. It does, however, seem characteristic of our time that intelligence has begun to sever itself from traditional, conventional and social prejudices. It has been discovered that this breakaway makes intelligence considerably more effective, and therefore intelligence has increasingly established itself, not only in the sciences but in the general attitude of men. We see this exemplified at present in the use of physical violence in intellectual protest in so-called students' revolts, and in the arts where originality is held higher than aesthetic or social values.

It is one of the fundamental features of early child training that a developing child grows up into the traditional and conventional values of his own society. This process can fail in child development, as we have seen. Hence, consideration must be made of the manner in which so-called normal motivation and socially and traditionally accepted values develop in early childhood.

Motivational development does not flow like the development of consciousness or of motoric capacity from intrinsic qualities in the human being. Motor-speech and intellectual development are largely pre-determined in the genetic constitution of a new-born infant. Equally, it appears that individual consciousness evolves from the spiritual-historic make-up of the human race. Although we can probably assume that morality and conscience are also inherent as potentials in the spiritual constitution of a person, it is obvious that his actual code of morality derives from the influence of his environmental culture.

Thus, environment has a more basic and decisive influence on a child's motivation than any other aspect of his development.

On the surface, it may well appear that moral codes are acquired in early infancy through simple learning. Re-enforced by rewards and punishments, certain modes of behaviour and certain standards are adopted by a child, but simple learning situations do not lead to the acquisition of what is so essential to the establishment of moral values, since these are held more existentially and profoundly, as will be shown.

A more helpful image aiding an understanding of the mode of acquisition of moral values in a child is presented by the idea of identification. By this I do not mean the secondary act of identification of a child with *another* person, but rather the primordial experience of "oneness" before separation and alienation are completed.

In order to understand this, we need only remember what was described above as the infant's initial mode of experience. Once this initial Oneness of experience in an individual's life is accepted, we shall realise the tremendous importance that the mother, and later on the entire environment, must play in what we call a child's moral development.

The process by which (during the early infantile experience of totality and omnipotence) the Other—that which is not the "I"— emerges must of a necessity be an experience of frustration and disappointment. The first duty of a mother and later of a teacher is to ensure that the frustrations in which the child must become engaged as an inescapable and necessary part of his developmental needs are meaningful, in the sense that a child must find himself and his environment participant in a greater, unified and coherent world.

Although motivational development has its roots in earliest infancy, it proceeds and must continue through infancy and childhood, even through adolescence into adulthood. The main theme and argument in the motivational guidance of a child must be the concept of "The Good" in early infancy, "The Beautiful" in school-age, and only for a child growing through puberty into adolescence does the argument of "Truth" provide the natural cogent stimulation for his motivational development. A premature introduction of the *intellectual* argument, the argument of the *reasonable*, is detrimental to the establishment of moral values. There is a widely-spread trend towards this in modern education and it represents a danger to the motivational development of children.

III

Aspects of Developmental Handicaps

THIS CHAPTER will speak of children suffering from different forms of developmental handicap.

If one lets the totality and the variety of medical diagnoses and educational assessments of child handicap pass before one's mind, one feels confronted with a dismal, depressing and almost macabre picture. In contrast to this, I hope to show that the aggregate of developmental handicaps in childhood presents a wide, fascinating and beautiful panorama of human existence, exaggerated it may be, but being exaggerated, it is that much more revealing.

I do not intend to give objective and scientific descriptions of handicap, but I shall attempt by means of empathy, the power of putting oneself inside the experience of the other, rather to depict a particular disturbance or handicap.

Developmental failures should not be seen as isolated conditions, but as phenomena which can combine with others in various ways. Indeed, that which is commonly regarded as normal development is nothing other than a relatively harmonious balance between the many possible and different aberrations and failures to which development is exposed.

1. Morning and Evening

In young children, the head is always larger relative to the body than it is in adults. In some children, this difference is more marked. It may occur that a child's head begins to grow noticeably in the first weeks or months after his birth, or growth will be less noticeable because more gradual. We see nothing unusual in a child's particularly large, and more often than not, beautifully-shaped head, nor need there be much evidence of largeness in a measurement of its circumference. Evidence of this feature perhaps lies more in a particular formation, e.g. a domed brow, or wider temples, but once we have become accustomed to the general run of shapes and sizes of children's heads, we have little difficulty in discerning large-headedness or its opposite in children.

One can see that a typically large-headed child (in whom the head is noticeably large in proportion to his body) usually has small, fine hands and particularly small feet. Generally, in these children, the limbs seem finer and less developed than usual.

Large-headed children do not only show typical physical characteristics, but emotional and mental ones as well. Very often, they not only look like little princes, but they have a certain prince-like aloofness and gentility in their manner. They are inclined to be dreamy and somewhat withdrawn. They tend to pursue their own thoughts and fantasies, and take relatively little part in rougher play-activities of their contemporaries.

As a rule, they will have learned to speak articulately at an unusually early age. Speech often develops before standing or walking and, in fact, movement-development in general seems to be a little reticent or delayed. These children are inclined to remain passive onlookers.

Their development of language is striking, and even when they grow up in areas in which less refined dialects are spoken, they will develop a fastidious, careful mode of speech of their own, so that it is often not easy to trace their cultural background from their language alone.

Large-headed children tend to be irritable and over-sensitive, often unable to concentrate or to apply themselves for any length of time. While still toddlers, they may pass for ordinary, healthy children if emotional difficulties do not arise in the families, but they often begin to cause concern when they go to school. The teachers usually complain of inattentiveness, dreaminess, irritability, of lack of concentration, and on occasion, aggression and obstreperousness in these children.

When large-headed children are not met with understanding, considerable and serious problems can develop. They will present a picture reminiscent of severe emotional, even psychotic disturbance. They will incline to lugubrious fantasies, to pre-occupation with death and the like. They will have phases of swearing as well as moments of totally unreasonable and uncontrollable aggression and violence.

In later childhood and adolescence, the limbs of large-headed children may accelerate in growth, but the size and particularly the shape and proportions of their heads remain salient and with this, a tendency to serious maladjustment persists with all its possible consequences, if the basic condition is not noticed and understood.

It is helpful to have encountered the extreme form of this condition, which is hydrocephaly—a medical pathological condition in which the internal water in an infant's head burgeons to such an extent that increasing pressure not only enlarges the skull, but to a degree diminishes the brain-substance. The head can grow to such gigantic proportions that it cannot be raised or held up, whereas the body remains diminutive.

Children suffering from this condition in an extreme form which often leads to blindness and complete physical incapacity, are known to have spoken concisely and clearly before they were overtaken by death caused by ever-increasing pressure in the head.

It is interesting to consider the head as such, as it is in early infancy and even before birth. It is impressive that immediately before birth, the whole child takes up a specific position in the womb and has an outline which is identical with the outline of a developed human brain. In a sense, one may say that we are all born as a head. The head of a new-born infant is nearly as large as the rest of the body and nearly equal in weight.

It is an interesting and telling fact that mothers of large-headed children often report that they felt particularly happy throughout pregnancy, so much so that they were reluctant to part with the child when the time came to do so.

This observation is of specific importance, because large-headedness is an expression of these children's own reluctance to be born. They seemingly would retain the pre-natal proportions of head to body. The process of descent we have described in the last chapter as basic to child-development tends to be hesitant and confined to the head itself. Features associated with the head such as language and the power of fantasy are developed to a high degree in these children, whereas bodily skills and abilities are largely neglected.

Against a developmental background, a large-headed child is prominently one who wishes to remain a child in a child's world of fantasy, rather than grow into reality, fact and the world of adulthood.

The head is that part of the organism that senses. It is carried and served by the body. It has to be transported, like a king or minister of state in his carriage. It has to be supplied with blood and oxygen by the body. The head is the observer, planner, organiser, but not the doer or the mover. It is truly the prince, the aristocrat of the bodily organisation.

The large-headed child must be allowed his princely nature. He cannot help being the one who gives the orders because his nature is so predominantly the imperious one of the head. When orders are given to him and he is expected to work and to carry them out, we so frustrate him that he is driven to despair, from which he takes flight in unfortunate fantasies and aggression.

It must be imagined that to fulfil a demand for work is to these children as preposterous and silly as to us, if we were expected to walk with our jaws, or saw wood with our teeth, or to dig a garden with our noses.

It is an unfortunate fact that many adults react to the aloofness of a large-headed child with frustration. They think that after all a child is a child and must obey. He must do what an adult says, which presupposes that the relationships between adult and child can be likened to the relationship between head and body. In the case of this particular type of child, the relationship is rather the reverse and may be more that of the servant to the prince.

Hence, the first approach to the large-headed child should be one of tactful acceptance of and concession to the superiority which he feels himself to possess. It must be compassionately understood that it is the unavoidable consequence of his developmental situation, and it then can be seen what can be done to help him.

Above all, the large-headed child must not be taxed with repetitive modes of learning, but things must be presented to him in the teaching situation in a concise and compressed form so that he can grasp them at a glance and form concepts without having to strain his memory. He should be allowed to exert his fantasy and make use of it, and of his ready grasp of symbol and meaning, of his powers of abstraction.

The teacher must relinquish the wish to be adult and superior, to enforce the teacher-to-child relationship in which the child appears as the pupil, as an inferior, as one who must obey. Once we have established the correct therapeutic relationship to the large-headed child and in doing so, have won his confidence, we can attempt to help him to overcome the bias of his constitution.

This entails helping him to achieve a further descent into his own body. His sensitivity must be coaxed into his limbs and their extremities, his fingers and toes. The following, for example, can be done. In a game-situation, a number of objects can be hidden under a cloth and a child encouraged to discern their nature by feeling them under the cloth with his fingers. Then, as a second step in therapeutic training, with his toes. With an older school child, it helps to get him to write with a pencil held in the toes. By means of various similar exercises, motor-descent into the distal parts of the body can be stimulated and supported.

Educationally, apart from making positive use of his power of fantasy through "nourishing" it with tales and legends appropriate to his age, a large-headed child's power of language and his symbolising faculty should be carefully guided. Abstract mathematical thought is to the large-headed child quite accessible, if it is not introduced on the basis of memory and repetition, but of spontaneous grasp. Naturally, steps in learning must be appropriate to a child's age and small enough for that child to take successfully. Under correct guidance, these children can make astonishing progress.

It is essential that teachers persevere in maintaining the therapeutic attitude described above as imperative in dealing with the large-headed child, otherwise in respect of gaining these children's interest and co-operation, they will be defeated at the outset or can lose it at any subsequent time.

It may be helpful to think of a situation in which we ourselves might have had a "large-headed" experience. Are there occasions in ordinary life when a person reacts rather with his head than with his body, which remains relatively ineffective? When is the head over-sensitive? When is a person easily irritated, because he hears and senses things acutely, but cannot execute the demand they impose upon him?

It is likely that most people feel precisely like this upon waking in the early morning. One wakes up into one's senses relatively easily and can become alert to the degree of irritability, but one has difficulty in gaining control of one's limbs.

Inasmuch as we can relate ourselves to this peculiar early morning state which varies in different people and at varying times of life, we have gained some intuitive awareness of what it is like to be a large-headed child.

Thus he can be thought of as a "morning child", and in an artistic appreciation of the qualities of the early morning—of dawn, when the colours have not yet become distinct, when mist envelops everything and fairy-tale, magic and myth have not yet receded—we are allowed to enter, as it were, the "territory" or the "landscape" of a large-headed child.

At the other end of the scale, there are children whose heads appear small, with foreheads narrow and receding, noses usually well-developed, chins pronounced, and whose limbs are big and strong, with large hands and feet, all with a promise of growth and strength.

Small-headed children are usually very outgoing, helpful, willing to serve and do the things that others plan. Their movements may be slow and clumsy, but they are hard workers, industrious, and once they have started something, they will plod along with it. They are realists, for the most part devoid of imagination and fantasy. When confronted as children, with Father Christmas they will be inclined to wonder why their neighbour, Mr. Jones, is wearing red trousers and a beard. They have extraordinary difficulty in grasping the meaning of abstract concepts, are slow to learn, form and symbol seem inaccessible, and although they are willing, they are easily frustrated and dis-illusioned.

In more severe cases, speech in small-headed children can be laboured; they may be inclined to dribble, articulation of speech is poor, and they may be impoverished in linguistic expressions.

The small-headed child has to face considerable trials in his scholastic pursuits, but once he has grasped a thing, he wants to put it into practice straightaway. Although grasp and recognition are slow in these children, they remember well, which allows them to learn by repetition and to acquire skills through continued training. When slowness and limitations in these children are misunderstood and interpreted as laziness, when demands are made on their intellectual faculties that they cannot meet, they are easily frustrated and develop feelings of inferiority and failure of a serious order. They become sullen, stubborn and unco-operative as a result.

Once a small-headed child has been driven into a defensive attitude, it is difficult to help him to recover a more co-operative and positive frame of mind. Unlike a large-headed child who, even when he has become severely maladjusted, will respond rather readily to understanding and skilful handling, a small-headed child is inclined to be as tenacious in his despair and maladjustment as he is in his good-will and perseverance.

While an excellent beginning of a therapeutic, constructive attitude to small-headed children lies in the acceptance of their specific limitation, due note should be taken and remedial use made of their willingness to serve and do things preplanned for them.

Educationally, they should not be presented with intellectual problems which have to be grasped from the outset, but use should be made of their readiness to learn through repetition and memory. Learning must proceed along practical lines. A letter of the alphabet, for instance, drawn on the blackboard or worse, in an exercise book, may prove incomprehensible in spite of years of arduous work to copy it and grasp its significance. But when, for example, the letter is drawn on the floor in a large form and the child can "walk" it and absorb it into his whole body, he will gradually become capable of raising the form up into his perceptual consciousness.

The entire therapeutic approach to the small-headed child should always be aimed at raising and refining larger basic movements into visual, conceptual form. While a large-headed child has to be led from the word and concept to experience in feeling and from there into the sphere where things are actually *done*, precisely the reverse must be attempted with a small-headed child. Whatever he has to learn. his learning must derive from a practical sphere. It can be raised progressively to the state of emotional appreciation and ultimately into the abstract sphere of word and concept.

In a small-headed child, may be recognised a too intense, too early descent into his body and limbs with a resulting reluctance to develop speech and a lagging behind in language, so that often, speech comes only in the third, fourth or even fifth year.

Are there situations in which an ordinary person like oneself reacts in a "small-headed" manner, in which one's limbs are willing to continue what they are doing, but one's head, one's consciousness can no longer take in new stimuli, is no longer willing to interpret or to grasp things? Is this not the condition of a person of an evening after a hard day's work?

The evening is a long way from early morning mist, from the promise of dawn. Colours have distilled into clarity, the horizon appears in greater distinctness and things stand out in their reality. One faces west, towards the setting sun. At such times, we are in the situation of a small-headed child, and thus we may justifiably call him an "evening child".

Large-headedness or small-headedness, if not extreme, need not constitute handicaps in themselves. Yet in more extreme forms, they can present very serious developmental problems. These two extremes were first described because they most clearly demonstrate developmental pathology to be exaggerations of a normal human constitution. Large-headedness and small-headedness are elements of human make-up, between which anthropological polarities must be struck, although we all sway between the two extremes in the course of time, minutely and in fine measure.

Therefore, the phenomenological study of the developmental polarity between head and body, above and below, and between buoyancy and weight is basic to an empathetic appreciation of developmental failure in childhood. Elements of large-headedness and small-headedness are always involved. Every handicapped child in particular should be seen in the light of this polarity, because the remedial approach is relatively simple.

Later on, I hope to show that certain types of handicap and disturbance are linked to large-headedness and others to small-headedness. But even in approach to a normal, healthy child and particularly to a school-age child, it is well worthwhile to take this polarity into consideration, as many problems for the *normal* child can thereby be understood and successfully met.

2. Left and Right

The foregoing descriptions of large-headedness and small-headedness in children were based on child development as seen from a point of descent, from above to below. This direction is, however, merely one of the three fundamental dimensions of space. Besides the vertical dimension, there are those of depth, of back to front, and of lateral extension from left to right. In terms of human development, these dimensions are dissimilar in character.

Both the vertical and the saggital—back to front—are based on polarities. Head and body, face and rear stand in asymmetrical relation and are polar, but laterally, the organism is basically symmetrical, the left side of the body being a mirror-image of the right, even though there are slight differences between the two sides of a human form.

Again by means of observation and imaginative, interpretative empathy, an attempt may be made to see how overt elements of form can be expressive of specific developmental problems, be they intellectual, emotional or moral.

The left–right problem contains its own polarity which is one of *symmetry* or laterality on one hand, and *"dominance"* on the other. Laterality, or symmetry, is the element of form in the left–right phenomenon, not only of the finished form of the human body, but equally of its development, while dominance is an element of function.

In its embryonic state, a human organism develops, particularly where the face is concerned, in a bilateral symmetrical fashion, being formed out of distinct left and right halves which only gradually fuse at a median line. In fact, it is as though *limbs* would converge from left and right, not unlike arms, to join ultimately in front to form a face. Mouth, lips and nose in particular are formed from these "limb" parts fusing in a mid-line. All mid-line defects such as hare-lip or cleft palate, and certain defects of eye-formation, belong to the incomplete-laterality-fusion class. There are other defects deriving from this particular pathology like hypertellorism in which a person's eyes are rather far apart and the bridge of his nose is particularly broad, although no cleft or opening shows.

Twins, particularly identical twins and so-called Siamese twins, are born as a result of a specific order of progress in embryo of laterality

47

and fusion, but since this is primarily of a morphological nature, it shall not be dealt with here, and only the functional aspect of laterality and dominance will occupy our attention.

Before these questions are taken up, however, it will be necessary to gain a preliminary, qualitative appreciation of left and right as such.

To begin with, it may be helpful to understand that although the body is morphologically symmetrical, functionally with regard to the inner organs, this is not the case. The brain and the blood-vessel system are largely symmetrical. Symmetry extends even down into the neck. In the middle system of the chest, the body begins to lose its total symmetry in respect of its functional organs. The lungs which are still by and large symmetrical have on the right side three lobes and on the left side, two. The heart is not centrally situated and is not a symmetrical organ in itself, its left ventrical being very much larger and more powerful than the right, and the whole heart is placed not only rather near to the left but in a "free" direction from the upper right rear to the lower left front so that it lies on a skew diagonal, across the symmetry of the body.

From the diaphragm downwards, the inner organs completely lose the quality of symmetry. The liver is found on the one side, the spleen on the other. The liver extends well across the middle from the right side, and the spleen takes up only a very small volume of the far left. The stomach is placed to the left and the gall-bladder to the right, so that acid and alkali are laterally distributed.

Thus a basic qualitative appreciation of the fact that the body is *morphologically* bilateral and symmetrical, whereas it is *functionally* polydirectional and asymmetrical is arrived at. Human beings have this in common with all mammals, but unlike animals, we qualify our form-symmetry through Dominance.

Normally, each one of us uses one hand in preference to the other. This hand is more clever, more able to execute finer skills. It is used for writing and for all more differentiated and complicated movements. There is also a bias towards the use of one ear all the time and listening or hearing is done more acutely with it than with the other. Something similar applies to the eyes. Aim is taken or a telescope squinted through invariably with the same eye. Likewise, the same foot will lead in jumping or climbing.

Preference for or the preponderance of a particular side goes under the title of dominance. The great majority of people are right-handed

48

in their dominance, which is independent, or at least can be independent, of the intrinsic strength or efficiency of the selected organ or limb, and the choice of side is closely linked to the development of speech and language.

Although the majority of people are right-handed, only just over half of the population is completely right-sided, which entails dominance of the right ear, right eye, right hand and right foot. Only a very small proportion of the population, less than 5 per cent, is completely dominant on the left. A fairly large percentage, just less than half of the population, has mixed dominance, but of these, the majority has a dominance of the right hand.

Dominance, which is peculiar to the human being, develops in the course of early childhood. It is known that dominance development is closely associated with language development, and its first signs appear in a young child at about the second year, in conjunction with the first manifestations of speech. Dominance, however, only becomes fully established in the course of early schooling, at a time corresponding with that of the acquisition of the ability to read and write, having receded into the background between the third and the fifth year.

The mechanism of dominance development seems to lie in the fact that lateralisation in the *brain* begins to take place in connection with speech and language development. This means that the two hemispheres of the brain, originally bilateral and symmetrical to each other, become differentiated through speech development into one leading and one subordinate hemisphere. Although morphological and genetic factors play their part, it is important that it be realised that ultimately, dominance development derives from speech and language development. The lateralisation of the brain is a *consequence* and not the *cause* of speech development.

The importance of this must be emphasised, because it shows that the idea, widely held today, that the manifestations of the soul and mind are the results of the anatomical and functional structure of the body is a one-sided one. Undoubtedly, bodily structures are the instrument of, and provide the "material" for, human development, but it must be understood that the use of this instrument by an individuality and the influences of environment play intense and overwhelming parts in development as such.

This would seem to be particularly the case in speech and language development, which, although it is based on intrinsic genetic potentials, has to be stimulated to attainment by the influence of a specific culture on a growing child.

Once lateralisation of the hemispheres of the brain is established, there is a tendency to resist further change. It is interesting to note, however, that until puberty, such changes are relatively easily accomplished.

After this theoretical introduction to the problem of dominance, the earlier question concerning the *qualitative* differentiation of left and right can now be approached. It will be understood that this differentiation will really be based on the difference between dominance and laterality, and that because the majority is right-handed, it is customary to link the right side with dominance and the left with recession or non-dominance. Thus, when the term "right" is employed hereafter, this is in reference to the dominant, active side, in most instances.

In this sense, right can be spoken of as "day" and left as "night". The right is associated with activity and doing, and the left with passivity and receiving. Right means attack; left defence. Right means reality; left signifies fantasy or imagination. Right is the sword; left is the shield.

Every child needs to develop these polar qualities in a decisive yet harmonious way. Pathology, therefore, does not lie in a child's being left-handed instead of right-handed. It lies in a child's not having developed uniform dominance. This can mean either that crossed-dominance has developed, in which dominance of ear, eye, hand and foot is discontinuous, or it can mean that there is a failure to develop dominance altogether, which will result in ambidexterity or double-lefthandedness.

It is an interesting fact that the conditions of crossed dominance and ambidexterity by themselves need not necessarily cause any handicap. On the contrary, there are and were great personalities such as Leonardo da Vinci, who drew with one hand and painted with the other and who could write with both hands in different directions and in mirror-writing at the same time, but this was and must be an exceptional, conscious achievement.

Very frequently when a child is otherwise handicapped, crossed dominance and ambidexterity can cause considerable developmental difficulties and problems. They indeed occasionally occur in perfectly normal and well-developed children and for this reason, it is essential that the question of dominance in childhood be considered in all cases.

From the developmental point of view, dominance of eye and hand is of greater significance than that of ear and foot. This is because of the overwhelming importance of reading and writing in our civilisation. These skills require that the upper right quadrant is established in the frontal field of attention. This means that writing proceeds from left to right and from top to bottom. For establishing the reading and writing direction, in the right upper quadrant, like dominance in eye and hand is of considerable importance. It is not impossible for a child who has a crossed dominance between eye and hand—between, for example, left eye and right hand or vice versa—to learn to read and write, but very often, even normally developed and highly intelligent children encounter considerable troubles in learning to read and write because of a crossed dominance between hand and eye.

There can sometimes be an absolute failure to acquire reading and writing abilities, or a child will use mirror-writing, or he will be obliged to battle with English while he masters arithmetic and other disciplines with ease.

Interest in the study of crossed dominance was greatly stimulated when one of the most promising students at an American college delivered a paper written in indecipherable hieroglyphics. The professor, suspecting a hoax, asked the student to read the paper to the class, which the student did with great eloquence. The student then explained that he had never been able to write in the usual manner, and it was discovered that he had crossed dominance of hand and eye. It took this student seven years of intense therapeutic effort to adjust his dominance and to learn to write.

Although for practical purposes, the dominance of eye and hand only is of real significance to a child's development, it is necessary for therapeutic reasons to make a complete assessment of a child's dominance before a decision as to the way help may be given can be reached.

It is not always easy, particularly with very young, or very handicapped and disturbed children, to make a reliable assessment of potential dominance. To the extent, however, that communication can be achieved or interest roused, the person attempting assessment will gradually be able to determine, for example, which is the dominant ear by getting a child to listen to a clock or watch. To determine which is the dominant eye, a hole can be cut in a piece of paper, which is then given to a child that he may look through it at the tester. A very young child, in whom dominance has not yet been established, will more often than not put his nose or mouth to the hole, but when dominance of eye has begun to establish itself, he will usually show clear preference for one specific eye when looking at one through the hole. Another means more successful with somewhat older children is to roll up a piece of paper into a "telescope" through which a child may look. In both cases, the paper or tube should be given into *both* hands of a child so that the choice of eye is not influenced by the dominant hand.

The dominant hand can be ascertained by giving the child a pencil to use or by observing which hand he uses for skilful movements in feeding, for instance. Some children may use one hand for tasks requiring greater strength and the other hand for finer movements. The latter will be the dominant hand.

The foot that is dominant can be revealed by getting a child to step up on to or jump off a chair, or to kick over a small object such as a matchbox set upright. The leading foot in each case will be dominant.

It is of importance, as has been said, to establish dominance in all these four things before deciding how to influence or alter a crossed dominance. Under no circumstances should an attempt be made to change a child's *main* dominance. Therefore, if a child is left-eared, left-handed and left-footed, but right-eyed, he should probably be retrained to become left-eyed as well, when the three other dominances are really distinct. Other preferences being equal in a child in whom there is a prediliction for the right, help should be given to establish right-sided dominance of both hand and eye.

When dominance is already firmly established in a hand, it is easier to shift dominance of eye, but before this is attempted, the eyes require to be tested for visual acuity. One must avoid training dominance into an eye which is not equal to the task.

All this is not intended to tempt parents to interfere with the spontaneous development of dominance in their children. This is best left to itself. Signs of possible dominance problems may show in speech development or later on in school, when a child has trouble in learning to read or write, at which time, especially, he may reverse many of the letters, particularly b's, d's and p's, or leave out and misplace letters, or have difficulty with the ordering of letters when learning to read. A test of his dominance is indicated at such times and if crossed dominance is found, an attempt should be made to correct it.

The easiest method of correcting dominance of eye is to supply a spectacle frame which, on the side on which dominance is to be suppressed, contains a red glass. While wearing these spectacles, a child uses a red crayon or red pencil. This allows him to use his ordinary binocular vision for orientation, but he will see the red writing he is producing only with the eye that is not covered by the red glass.

As most reading material in school comes in black print, a child will require in addition to the red glass, a sheet of green cellophane—green being complimentary to red. If this green cellophane is placed over the black print of the book, the print becomes invisible in that it merges into a black field for the eye that is covered with the red glass and only the uncovered eye is able to discern the black print through the green cellophane. This double approach to dominance retraining of the eye is very helpful and usually adequate.

If dominance of *hand* is to be changed, the co-operation of the child must be sought and the continued support of the teachers obtained in making the decision to use the hand exclusively that is to become the dominant one, particularly in writing, though equally in all other complex and subtle activities. For the more alert child with crossed dominance, it is of considerable additional help for his learning to write to introduce him to mirror-writing in the four frontal quadrants. This is done by dividing a sheet of white paper into four quadrants by means of two lines, one horizontal and one vertical. The child is then shown how the same word can be written in mirror-form from the right upper quadrant which is the normal one for writing, to the left upper quadrant and from there down to the left lower quadrant and back over to the right lower quadrant, so that now the left lower quadrant presents the mirror-image of the one above it as well as of the one on the right.

TƆI9	PICT
⊥ƆI9	bICⱢ

For a period of a few weeks, the child should be encouraged to write a variety of words in all four quadrants in the mirror-fashion described. Then, the two lower quadrants should be omitted and the child should continue to practise writing words mirrored in the left–right direction for a number of weeks. Finally the left upper quadrant is also omitted and the child's writing guided into the upper right quadrant, which becomes the sole one to be used.

It is, of course, only too understandable that the proper establishing of dominance, which can relieve a child of very painful struggles in learning to read and write, has an influence on his emotional development, but it is not only the relief from scholastic handicap that works back on the emotional condition. It is likely that crossed dominance in itself generates a certain emotional uncertainty in a child.

The emotional aspect of dominance and laterality becomes very much more obvious in those who are ambidexterous. There are cases in which, in spite of normal language development, dominance is not developed in hand, eye, foot or ear. From a certain point of view, these children would seem to be at an advantage, when it comes to playing tennis or fencing or the like, for instance, because their complete ambidexterity should place their opponents at a disadvantage. Yet, ambidexterity often passes unnoticed, not only by parents and teachers, but even by the child who possesses it. An apparently unrelated problem occurs, in which a child of this ilk is accused of persistent lying and stealing, yet, he is a nice, lovable fellow with open, good contact. His parents and teachers have tried everything to combat his problem. They have given him continuous affection and support. They have also faced him with the consequences of his misdemeanours and subjected him to punishments of all kinds, all to no avail. The child continues to steal what he fancies wherever he finds it, and he will invariably lie in an inventive, blatant though transparent manner. In fact, he does not even seem to take any great trouble to make his lies credible. He simply tells a story and, moreover, will not hesitate to tell a different one when the same situation recurs, regardless of the effect it has on those who hear it. He does not appear to suffer from conscience at all. He will promise not to lie or steal any longer, but the first occasion which presents itself, he will do it all again, and the whole problem is again exactly as it was before.

Sometimes these children can be recognised at a glance. They have a physiognomy in which left and right halves of the face are identical.

In a peculiar way, their faces look round; possibly a little catlike, and when, for example, they are asked to write with both hands, they show exceptional ambidexterity often to their own astonishment.

If such children have failed to establish dominance and are still young enough, it is often possible to help them to establish a sense of ethical behaviour by training them towards a definite dominance. Their problem is simply that left and right have never differentiated, and with this, truth and falsehood, and right and wrong have remained vague and undetermined.

It needs months and sometimes years of continuous effort to help a child suffering from this condition to develop sufficient perseverance and willpower to establish clear and definite dominance. If, however, it can be done, it usually entails a triumph over the child's moral problems and a victory for himself. After puberty, unfortunately, this victory becomes well nigh impossible.

It must, of course, not be expected that every moral difficulty in a child is based on the failure to develop dominance. By no means every child who lies or takes things does so in the manner characteristic of an extremely ambidexterous child, who, as we have described, shows a lack of conscience or emotional participation. Very much more frequent in children are, of course, those moral problems that arise from other difficulties encountered in their lives, from emotional frustrations with which they are unable to deal, from disappointments they had to go through, or situations they were unable to solve. There are also organically-based conditions that can cause moral problems in children. However, the moral problems that arise as the result of a lack of dominance are fairly unmistakable, once they have been encountered.

Equally, it should not be thought that every failure in learning to read and write on the part of a school-child is caused by crossed dominance, but if a child's scholastic ability lags severely behind and does not correspond to his measured intelligence, and particularly if there are such indicators as the reversal of letters and the like, it is always worthwhile to investigate dominance development.

Again, the problem of dominance and laterality does not necessarily appear in isolation, but more usually as a contributory factor in a great variety of developmental handicaps.

3. The Palsied Child

The affliction of palsy has accompanied mankind from as far back as human history reaches, but only in the last century has the medical aetiology of palsy been established. Largely as a consequence of this advance, it has become rather difficult to regard a palsied child, a palsied person, as the total phenomenon he is and thereby arrive at a more intimate understanding and appreciation of the personal aspects and problems of cerebral palsy. It will, therefore, be helpful to recapitulate briefly that which was outlined in the first chapter of this book.

While the cause of cerebral palsy lies in brain damage sustained at birth, it does not immediately become visible. Some of the symptoms of cerebral palsy, such as the scissor-gait of spasticity and the involuntary movements of athetosis are both normal phenomena in the second and third months of child development.

It shall here be attempted to see cerebral palsy not from the point of view of the anatomical, aetiological consequence of brain damage, but rather from that of failure of early child development, which, though obvious chiefly in the pathology of motor or movement development in the limbs, finds a less obvious but much more basic, much wider and more existential expression elsewhere in a child's total being-in-the-world.

Firstly then, an evaluation of the basic symptoms of cerebral palsy is in order at this point. There are two points of view from which the symptoms of cerebral palsy can be delineated. One pertains to the qualitative differences in the nature of the paralysis, while the other is spatial, and relates to the parts of the body that are affected, particularly the limbs.

Paralysis is categorised into quadruplegia, paraplegia and hemiplegia, the latter referring to paralysis of one side only, either the left or the right, where both arm and leg on that side may be paralysed, or only one of either. In the case of paraplegia, either both arms or both legs are affected, equally or one more than the other. Finally, in quadruplegia, arms as well as legs are affected and although one limb is generally more severely affected, there is basically an affliction of all four limbs.

It will be readily appreciated that in any of these forms of paralysis, the spatial experience and spatial integration of a child are fundamentally involved. In hemiplegia, lateral orientation is impaired. In paraplegia, vertical orientation is inferior. In quadruplegia, orientation the sagittal dimension is deficient.

The latter is a particularly striking phenomenon. The disturbance of secure self-experience in the rear–fore orientation in a quadruplegic child is encountered in the expression of intense anxiety in his eyes, which indicates that he is completely imprisoned in the frontal direction.

With regard to the *qualitative* categories of paralysis, four separate things may be spoken of. Naturally, in very many cases—in the majority of cases in fact—these qualitative elements combine severally in the final form the paralysis takes. Nevertheless, the four qualitative elements can and should be distinguished.

First there is spasticity, in which a limb or joint is stiff, and the muscles contracted, but one can usually move the limb if this is done very slowly and gently. However, any increase in the speed of moving the limb, any abrupt pressure will cause increased stiffening and contraction in its muscles. The spastic condition can perhaps best be described as a state of constant and extreme muscular cramp and contraction. It is, therefore, the very opposite of "paralysis" in the exact sense of the word, as there is no lack of muscular activity, but rather a freezing into a sort of quasi-permanent over-contraction.

It should be understood that controlled movement does not consist only in the ability of the muscles to contract but equally in their ability to relax, and that movement in an intentional and co-ordinated sense is only possible when a complex of muscles co-operate harmoniously in their contraction and relaxation in obedience to a dominating plan or "theme", which may be called the "melody" of the movement. This very co-operation and co-ordination becomes impossible in the case of spasticity, because of the constant, total contraction of the muscles.

The second form of palsy, athetosis, consists in involuntary, often twisted or spiral, worm or plant-like movements. The attactic form is characterised by over-shooting movements. A response to an intention results in muscular contractions which are appropriate enough, but these are insufficiently compensated by opposing muscles, so that the consequent movement over-shoots its aim. Finally, the rigid form of palsy allows for only the slowest of movement, and when the limbs are so moved, it is as though against a lead-like resistance.

It is interesting to note that none of these four symptoms strictly conforms to what is meant by paralysis, which implies the inability to move altogether. It can be seen, however, that the symptoms of cerebral palsy are pathologies of *movement*, in which movement is more often distorted by over-intense muscular *activity* rather than by its complete absence.

Complete flaccidity, the inability to bring about any muscular movement—the so-called flaccid palsy—is typical for *peripheral nerve lesions*, in the spinal cord or the limbs, but does not occur in the syndrome of *cerebral* palsy.

Another important observation for the evaluation of cerebral palsy is the following. Cerebral palsy does not only impair the child's motor-development and thus delay or render impossible things like sitting, walking and other free, co-ordinated movements, but it also often results in distortions of the human form through sustained contraction. Although mothers and nurses are usually aware of the fact, teachers often may not realise that the child they see in a contorted, disfigured position, with wrists and hands severely over-flexed, legs crossed, and feet and toes extended, can regain his harmonious and proper human form the moment he goes to sleep and relaxes completely. In the morning upon waking only, the contractions and distortions reappear. To the ordinary spectator, even the most severely palsied child (at an age when the contractions have not yet permanently distorted the frame) during sleep appears normally formed and perfectly healthy.

From these descriptions, it will be understood that cerebral palsy is by no means only a motor-disability. There is a very considerable sensory content to this condition, which consists by and large of a marked degree of over-sensitivity.

It is easy to observe how a spastic child reacts with increased muscular contraction to any abrupt noise, or even to the approach of someone who does not enter his field of vision gradually. I have seen cramp increase in intensity in many young spastic children as a shaft of sunlight from a window moved across their faces. It is astonishing to watch the slightest sensory stimulus, if unexpected, causing increased spasticity.

The same things apply to an athetotic child, although in his case it would seem that it is rather emotional than sensory stimuli that set off involuntary movements. Such a child will set hands and head in violent rotational movement when he feels sympathy or affection welling up in himself towards another person. It is his life-of-feeling rather than that of his senses that triggers off his athetoid movements.

In ataxia, voluntary movement results in over-shooting that can be enough to upset the child's balance.

When the description in the previous chapter of motor-development in the child is remembered, being one of descent from the head down through the body into the limbs, one may now be better able to interpret spasticity. It seems that the differentiation between head and body which is essential if motor-development is to take place normally, does not occur in the appropriate manner in a palsied child. The human being is so organised that the head is a pole of rest, of tranquillity, so to speak, not primarily intended to partake of movement but to be susceptible to sensory impressions. Only the very refined movements of speech and of thinking are meant to take place in the head. The head is not party to large, sweeping movements of the limbs. In fact, the head is the resting pole to which large movement is to refer, and from which it is to spring. In the case of a spastic child, this primary differentiation does not proceed far enough. It appears that sensitivity, properly localised and concentrated in the head, is here spread out over the whole organism. The whole body is over-sensitive, and is, so to speak, too much of a head, and conversely with regard to movement, the head is insufficiently isolated or separated from the body, and is gripped by impulsive forms of movement. This is particularly so in the case of athetosis, but it is also present in other forms of cerebral palsy.

If this is seen and understood, certain approaches to the handling—and to some extent, to the therapy—of the cerebral palsied child are opened up. As to the handling of palsied children, it is necessary to bear in mind their intense hypersensitivity and avoid sudden, abrupt and intense sensory stimuli and, in the case of an athetoid child, excessive emotional stimuli. Various remedial exercises and physiotherapies have been worked out for the palsied child, a number of which take into consideration the problem of the separation of the head from the movement-patterns of the body, and these things are particularly helpful to the young palsied child.

However, physiotherapy and curative eurythmy, which are two important therapies, require medical advice and guidance and fall, therefore, outside the scope of this book.

Nevertheless, the benefits of a general approach to the palsied child will be considerably augmented if his personal situation and his experience of his self are understood and if his imprisonment in sensory experience is sensed.

Equally to remedial advantage is an appreciation of that sensitivity which should be restricted basically to face and head but which instead extends over his whole body. It will then be realised that touching these children's limbs requires that tact and tenderness with which a face would usually be touched. Similarly, if one has not entered the field of vision of such a child, one should not start, for instance, to push his wheelchair around without warning. One must make one's presence and intention known to him beforehand. In general, the approach must be gentle, soft and gradual. One must avoid evoking an over-sensitive reaction in his body and limbs, and one must try instead to draw it up as it were into his head where it rightfully belongs. One helpful means of doing this is to have the child learn to follow his own movements in a mirror. Through so doing, sensory sensitivity and movement may begin to become disentangled.

When empathy is developed for the condition of a palsied child as so far described it becomes apparent that he is not only over-sensitive but that he lacks the experience of self, of his body schemata that the healthy person derives from his senses of touch, movement, well-being and balance. These senses operate largely at or below the threshold of consciousness, yet they are crucial to physical security, well-being and effectiveness, and in a palsied child these basic experiences are not established.

In respect of the lack of separation of head-sense function and body function in palsied children, the phenomenon of so-called scissor-gait is particularly telling. The scissor-gait is characterised by crossing of the legs when walking or trying to walk and in it we can recognise easily something that should properly take place in the eyes. In the act of seeing, the axes of the individual eyes must cross or intersect so that the images perceived by the separate eyes can be fused into one coherent picture. With cerebral palsy, this tendency would seem to extend to the lower limbs and result in the well-known phenomenon of the scissor-gait originally described in Little's Syndrome.

Normally, walking is necessarily characterised by parallelity and not by intersection. Divergence as opposed to "intersection" in the gait can occasionally be observed where the symptom of ataxia is foremost. Such a child's movements are jerky and excessive in general, but when walking in particular he tends to throw his feet forward and outward. He walks as if he were straddling something, and a gait showing the very opposite of the tendency to cross the legs is apparent. These children frequently have difficulty in focusing their eyes and are prone to a variable outward squint, so that the axes of the eyes do not seem to intersect, or, at best, cannot sustain an intersection once achieved.

While the spastic and attactic forms of palsy are pertinent illustrations of one sort of disharmony in the proper distinction between head and body, athetoid palsy indicates yet another form. Athetosis is as a rule more marked in the arms and hands and possibly in the necks and heads of those afflicted, rather than in their legs, and it has been pointed out that emotional rather than sensory stimuli set off the athetoid movement. Thus, a breakdown in the differentiation between the head and the "middle" system of breathing and heart-beat—of arms and hands—is encountered here. It is as though the function of these middle systems were predominant and then emotional responses result in movement-patterns that proceed upwards and take possession of the head.

If then, the particular sort of scrutiny heretofore described is brought to bear on sufferers from cerebral palsy, it is possible to see amongst its various forms a specific mode of breakdown of the human organism's natural differentiation into the three domains of head, trunk or middle and limbs.

Need of assistance in bringing these domains into proper harmony is perhaps more apparent in the athetoid form of palsy than in the others, and an athetoid child can be helped to achieve emotional control in the sphere of movement if exercises are given which enable him to learn movement to very soft, rhythmical music. Simple hummed rhythmic melodies are particularly beneficial as an accompaniment to passive movement exercises to begin with, and later for active ones. In fact, a combined use of colour and music has a particularly relaxing and harmonising effect in general on cerebral palsied children, and a therapy has been developed here in which the movements of a eurythmist are projected on to a screen in coloured shadows to the accompaniment of suitable music. In watching children, clear signs of relaxation and harmonisation can be observed.

In both the general and the educational approach to a cerebral palsied child, it must be borne in mind that he needs reassurance in respect of his spatial orientation and that he must be aided in building up his experience of his own body-image, all of which can be done in a variety of ways. There is another condition, very different from the cerebral palsies, which can severely impede a child's motor-development. This condition is usually called hypotonia. The aetiology of this condition is not known precisely, but in many cases it too would seem to be of cerebral origin. A child suffering from hypotonia does not begin to raise his head or to use his arms and hands at the usual times, but remains lying in a prone position. Sometimes, he has not acquired the upright position by the time he is of school age. The condition can combine with general developmental retardation and can be fairly severe. There are occasional symptoms of slight spasticity, but the main symptom is the opposite of spasticity, consisting of a rather low and flaccid muscle tone, general inertia and flaccidity. The child may otherwise appear to be relatively normal, or possibly a little sleepy and apathetic.

The approach to children with hypotonia must be a completely different one from the approach to cerebral palsied children. They will require continuous and usually fairly intense stimulation if motor and general development is to proceed at all. Frequent and continued excitation of the primary reflex of gripping as well as the foot reflex is essential, one helpful means of doing the latter being to get the child to try to stand with bare feet on a round stick. Only through perseverance in stimulation both of motor-activity as well as of speech will such children be helped, but again, such physiotherapy must be applied under medical supervision.

A further gross motor-handicap can come about as the after-effect of poliomyelitis which, however, is not a developmental handi-cap, strictly speaking. Poliomyelitis results in an interruption in the function of the nerves and severe flaccid paralysis can be the conse-quence. In many cases, a child who has sustained serious paralysis in one of his limbs will tend to suffer from emotional handicaps in the course of his development. Here, the understanding of the qualitative differences between left and right, above and below, will be of help in the endeavour to grasp the psychological consequences of this affliction. In more severe cases of poliomyelitis, the paralysed limb may remain behind in growth-development and gradually become a kind of lifeless and passive appendage. If the affliction is only on the left side, which is qualitatively the side of defence and receptivity, a child may react to this incapacitation by compensating with extreme aggression. Intolerance, aggressiveness and spitefulness can be the result of a severe left-sided flaccid paralysis.

Conversely, when the paralysis affects his right side, a child may react by being over-passive, over-indulgent and unable to concentrate or to develop sufficient initiative. An interesting phenomenon can sometimes be observed in a child paralysed in both legs, and where there is a compensatory development of strength in the shoulders, arms and hands with, at the same time, increased willpower and perseverance. It is as though the willpower and strength normally at work in the lower limbs were thrust up into the upper part of the organism, to manifest there both physically and in the development of the personality.

Far more frequent, however, than these major, dramatic failures in motor-development, are minor, and in fact, sometimes minute motor-handicaps caused by what is now called minimal brain damage. These may go unnoticed, as motor-control apparently develops fairly normally. Sometimes, the motor-milestones are slightly delayed, but not enough to be obvious, and yet by means of finer methods of investigation, it becomes clear that motor-co-ordination is limited and impeded in certain ways. The handicap may only become noticeable when an afflicted child tries to imitate rapid alternating movements with his hands or fingers. He will not be able to perform the movements with the speed and co-ordination appropriate to his age, and in some cases, the attempt will make him feel sick and faint.

Such slight motor handicap will not only be a hinderance in a child's ability to learn to write and carry out other tasks that require finer motor-co-ordination, but it will also impede his general development in various ways. One is usually insufficiently aware of how much the proper integration of one's personality and one's inner security are derived from the efficient functioning of one's body and motor-co-ordination. It has been seen in a previous chapter, for instance, that the development of speech and even the ability to think depend on the development of finer movement-co-ordination.

It requires and deserves a fair degree of insight and patience to help a child who has difficulty in motor-co-ordination to increase his motor-control and to acquire the motor-skill he needs in order to mature as a person.

4. The Restless Child

The hyperkinetic syndrome in childhood, as it is often called, is represented fairly by the following description:

"The instant Alec could move and run freely on his little legs, he did so incessantly. There was never a moment of peace, not an iota of rest, and this has gone on until the present. He is on the go at all times, moving about restlessly, taking things, handling things, knocking things off the table or pulling down the curtains. He has snatched most important papers off his father's desk suddenly and thrown them into the fire. At the slightest frustration, he throws himself on the floor and kicks and screams. With practically no provocation at all, he bangs his head, screams or bites his wrists. Even when anything fully captures his interest, he still moves both his hands in front of his face in an erratic, intense, cramped fashion. If he is not constantly watched or even physically restrained, he can run out into the street and disappear. Day or night, he may cry desperately for long periods for no apparent reason. At other times, he sits and rocks forwards and backwards or from side to side, but usually he does not sit at all. He throws the table over, refuses to eat and from the table he thrusts his full plate to break on the floor. He can pull the tablecloth off with everything on it. He seems incapable of adhering to the truth and readily denies his misdeeds, and often deliberately tells 'stories'. He does not seem to have any sense of propriety. He will take anything belonging to anybody, such as other children's toys, or personal things from his mother and father."

This brief and drastic description could well be that of a very severely developmentally disturbed child, and yet, some mothers will recognise in it the grossly exaggerated story of their own perfectly normal, lovable, delightful two-year-old child, and this realisation is of the greatest importance for the understanding of the condition. It is, unfortunately, usually lacking. This theme will be developed after the symptoms and the aetiology have been described.

The symptoms which, in their pathological form, are probably amongst the most exasperating and trying that can occur in child development, are often caused by an inflammatory process in the brain in early infancy, the so-called encephalitis, so that the condition has even been called the Post-encephalitic Syndrome. I mention this here, because that very different symptom of early failure in childhood development, encountered in cerebral palsied or "silenced" child, is also caused by an early impact on the brain. Typically it is, however, a very different type of impact from encephalitis, namely that of anoxaemia, which is a lack of oxygen supply to the very tender and fast-growing brain, of the foetus or infant, usually associated with a constriction in the blood supply. The pathology of the encephalitic impact is practically the very opposite of this. It is an inflammatory process in which, by the nature of the inflammation, blood rushes with great intensity into the brain and floods it with its warmth and metabolic powers. It may well be that this polarity in the aetiological processes that reveals itself when the two forms of impact are looked at imaginatively is of some significance to the polarity of symptoms, even from the developmental point of view. The restless, hyperkinetic, violent child seems in many ways to occupy the opposite extreme from the paralysed, frozen, isolated and silenced cerebral-palsied child.

The development of speech in a hyperkinetic child can be normal, or disturbed in various ways. The speech of some can be particularly loud and uninhibited, and can be limited or entirely absent in others. The larger movements of the body are usually well-developed, but some impairment of the finer motor-co-ordination as an expression of minimal brain damage can more often than not be found.

Convulsive disorders in the form of *petit-mal* or epileptiform seizures are not uncommon as expressions of brain damage in these children. A further noticeable feature in some more severely-disturbed, hyperkinetic, young children is rocking. These children sway backwards or forwards or sideways from the trunk, either in a standing or sitting position, sometimes crouching on knees and elbows, or striking the mattress or floor with the forehead. Yet another and odder symptom can be observed fairly often, that of over-breathing. These children will have phases in which they inhale rapidly and excessively for considerable stretches of time, and generally breathing seems to be over-intense throughout the day. Their rhythm of sleep is often severely disturbed—some cannot get off to sleep in the evening, and some wake up in the middle of the night, often with nightmares, which creates in turn a situation of strain in the family. In younger children with this kind of affliction there may be serious feeding problems, fussiness, peculiar predilections to bizarre habits, and food-fads continuing into youth. Masturbation can become excessive and other modes of sense-stimulation can take on near-addictive proportions.

Children may have small articles such as a piece of string or a stick which they dangle in front of their eyes or use to hit themselves. They may use stereotyped movements of fingers in front of the face to produce special effects of light. They may play with water, splashing with their hands or taking it into the mouth, at which point peculiar exercises of spitting and the like appear to take on special meaning. Some go through phases when certain taste stimuli become imperative —those of sugar and sweets or vinegar and other very savoury spices— and they can get up during the night to seek satisfaction from one or another of these substances.

We shall shortly return to the basic symptoms of the hyperkinetic syndrome in the attempt to interpret them developmentally, but before doing so, let us consider for a moment the aetiological point of view. At the root of most cases of hyperkinesis there is an early encephalitic incident, which can be a consequence of childhood diseases such as measles, chickenpox, mumps, rubella and whooping-cough, but which can also occur as a virus infection by itself. There are, unfortunately, cases in which the inflammatory condition of encephalitis is caused by immunisation or vaccination. Although the latter is infinitesimally rare in proportion to the numbers of children immunised and vaccinated, among children suffering from the hyper-kinetic syndrome as a result of an early encephalitic incident, post-immunisation or vaccinal encephalitis certainly does account for a small but definite proportion.

One of the most important realisations in connection with its aetiology, however, is that encephalitis produces completely different pictures according to the time of onset of the illness. Three types of post-encephalitic condition can be discerned according to the age at which the encephalitis occurred. When the illness occurs in early infancy and before school-age, the syndrome is typically that of the hyperkinetic, or restless, over-active, uncontrollable child. This picture can be complicated by a convulsive disorder which has, as a result, secondary gross retardation. The level of intelligence, too, will be impaired.

66

When encephalitis occurs in a child of school age, particularly in the early period between the fifth and ninth years, the picture is one of moral disturbance. Intellectual capacity and general development seem less impaired, but there appears to be a severe injury to the moral faculties.

When encephalitis occurs in adulthood, the after-effect is usually the opposite to that of early infancy and there is rather a gradual slowing-up of motor-ability, leading to what is known as Parkinson's Syndrome, characterised by a mask-like face without expression, decelerated tripping gait, shaking hands, over-all inertia and lack of motoric drive. One of the main complaints, voiced perhaps more in the past than nowadays, concerning hyperkinetic children of school age is that they have a particularly vicious and negativistic attitude, that they perform destructive acts with so much speed, skill and cunning that it does not seem possible that they could do them without premeditation. They will do the very things that hurt most, destroy those things that are irreplaceable or break those things that are of the greatest value, either materially or personally. Equally, their characteristics of deceit, lying or pretence are often described as being particularly calculated and vicious.

Although the intellectual abilities of hyperkinetic children need not necessarily be *severely* impaired, there is nonetheless always some degree of impairment, often considerable, and to interpret their anti-social behaviour as being premeditated and calculated is an inept thing to do, since they almost never have the means. Nevertheless, very understandably, they generate problems that are taken to be indicative of deliquency, and these are particularly serious in adolescence.

The degree of intellectual impairment by no means goes hand in hand with the degree of hyper-activity and moral disturbance. In some children, the intellectual capacity seems relatively intact whilst moral development has not proceeded beyond the infantile phases of childhood, and even in adolescence these children are delivered to their impulses, drives and desires as a person normally is only in infancy. But there are different situations in which intellectual development can be so severely impaired that speech has not been acquired, and there are other gross and severe limitations. There are examples of every gradation between these two extremes and it must be admitted that no *specific* intellectual functioning or disfunctioning is pertinent to this particular developmental failure.

Some hyperkinetic children, although aggressive and negativistic in their acts, establish direct and warm contact both to adults and to other children. On the other hand, some may suffer from a lack of contact and relationship to the degree of real and severe autism. Here, too, all gradations between these extremes are possible. Autism can be so outspoken that it is the predominant feature of the overall syndrome and the child must be regarded as genuinely autistic, but again, like intellectual malfunction, disturbance in contact is not specific to the hyperkinetic syndrome and we shall, therefore, not discuss it in this context.

After the discussion of aetiological, medical and psychological aspects of the hyperkinetic syndrome let us now return to the theme that arose earlier out of a mother's observation, who recognised the description of a hyperkinetic child as an exaggeration of her normal, healthy, lovable two-year-old. This can introduce the more general idea that these children present a picture that is reminiscent of that which is encountered in the normal development of children between two and five years of age, who go through a phase in which impulses tend to be intense, activities largely over-emphasised and performed with much energy and exuberance, and in which there is little uncertainty, little reflection. In this period, children have not yet developed the degree of social motivation, which will show in them later when they are at school. Although powers of imitation are vigorous at an early age, they do not result in the tendency to conform for which they will be ready only later.

A normal two to five year old child negotiates a phase in which an extraordinary motley of impressions and stimuli, as it were, rain down upon him He has not yet created a generalised image of experience nor a system of his own that can encompass the growing diversity of stimuli and percepts to which he is continuously subject. He learns to build all these into a meaningful and comprehensive whole gradually, but in the process of doing so, his life of fantasy and his experience of reality interplay in a dramatic, and as a rule, fairly incomprehensible fashion to the adult.

More particularly, a child of this age is proceeding towards the unique experience of his own Self, his own Ego, at a time when his ideas and concepts have not emerged to that degree which allow him to cope with what adults call the "world of reality". When one learns to feel oneself into the situation of a very young child, one experiences

68

something of the dionysian ecstasy called forth in mankind of earlier millennia when the experience of individual egohood charged their souls, an experience one may faintly savour in certain forms of intoxication, even alcoholic intoxication, with which some of the behaviour patterns we may observe in the hyperkinetic child are akin.

This comparison, like most comparisons, is somewhat one-sided and only holds good in some respects. The restless child's motoric ability is usually not impaired in any great way, and does not, as such, resemble that of the intoxicated or inebriated person. The comparison here lies rather in the realm of the experience of an over-intense, predominant and foreground feeling or Selfhood while drunk or "high". Regarding the hyperkinetic as well as the post-encephalitic child in this light provides a new insight into the restless child's modality-of-experience. There will be less inclination to think of him as vicious and cunning. A better ability will be acquired in sensing that his lack of control and inhibition flow from an accentuated force of sympathy and over-positivity rather than from negative feelings of viciousness and antipathy.

In fact, a better understanding is obtained, not only of the problems of the restless child, but of children and indeed of people generally, when we consider that, primarily, two forces are at work in the *psyche*, or emotional constitution. One of these is Sympathy, or love. This force persuades us to be outgoing, intentional, active. It, too, generates aggression. It is the basis of all human feelings, including love and hatred equally.

In opposition to the force of Sympathy stands the other primary force—Antipathy. By means of it, we reject, we hold things off, we are not outgoing and intentional, we are not determined by longings and desires, but rather we are detached by it from our experiences. This is the agent that enables us to form percepts, to recognise things, and in the end, to form judgments. Through Antipathy, conceptual and intellectual life can be developed in contrast to the emotional and intentional life that stems from Sympathy and aggression.

It appears that the point of equilibrium between these essential powers shifts in various phases of normal child development. They are phases of which some can be termed extrovert and others introvert. The balance between the two which at any time constitutes our behaviour, would seem to be disturbed in a very specific way in the hyperkinetic child, in that the force of Sympathy seems to drown the force of Antipathy which is reminiscent of encephalitis or as described earlier a process in which the brain, the organ of perception, is flooded by the blood, the carrier of metabolism.

It must be acknowledged that that force which detaches the person from his emotional experience and allows him to form concepts and judgments is particularly weak in these children, in comparison to the outgoing force of intentionality and sympathy. The therapeutic attitude to a restless, hyperkinetic child must therefore comprise of two approaches.

One must attempt to stabilise a child's hitherto little challenged and unbalanced force of sympathy and aggression. As long as it is believed that this force is negative, a stand shall inevitably be made against it. This not only frustrates the child through repeated restrictions and prohibitions, but those who handle the children themselves become increasingly irritated and frustrated. As soon as it is realised, however, that the force which drives a child to destructive and aggressive behaviour is one of sympathy and warmth, positive outlets and channels for this over-abundant and uncontrolled force in the child, can be discovered. It will gradually be understood that the only argument for this type of child is the argument of love. Arguments of reason, truth or aesthetic values will not reach the restless child.

It is interesting to note that the same is true for a normal child in his pre-school phase. He cannot be approached by any argument other than that of goodness and sympathy. He is willing to do anything for the sake of love, even though he is as yet completely impervious to arguments for right and wrong, or of truth.

Every mother knows how the normal, healthy restlessness and hyper-activity of the young pre-school child can be constructively absorbed into fetching and bringing, and that it is of the greatest delight for young children if they may fetch and carry the things their parents or others need for one kind of work or another.

The same principle should be made use of in the guidance of a hyperkinetic child. Possibilities must be discovered for him to exert his motoric energy and physical strength towards helping another person, so that he may express his outgoing interest and sympathy in a constructive fashion.

If the restless child is to devote his energies to others either in defending or protecting them, the right where necessary to defend or to protect his own territory or belongings must be conceded to him.

He cannot be expected to swing unreservedly to the other extreme, because it is equilibrium that is required. All these elements should be introduced and fostered in play as well as on any other level accessible to a particular child.

It is, of course, obvious that basic help cannot come from this side alone. There is the need to reinforce in a hyperkinetic child the weakened forces of detachment and warding-off that derive from the force of antipathy. The basic requirement here is not to expose him to too many and too rapidly changing stimuli. Important as it is to provide continuous new stimuli and experiences for other types of children, it is well nigh poison for a restless, hyperkinetic child to be exposed continuously to fresh stimuli. This is why it is so difficult to contain hyperkinetic children in ordinary schools and even more difficult in special schools, because most modern educational approaches are based on fairly intense stimulation, with the frequent use of visual and mechanical means. Useful as these means may be for many children, they are disastrous for hyperkinetic children.

A hyperkinetic child requires to be surrounded by a very consistent environment, and he should be well sheltered from overstimulation. Ideally, he should only encounter situations and stimulations with definite values and meanings which are strongly upheld by others in his surrounding. This means that the child need not go from situation to situation, from experience to experience, re-adapting on each occasion, but that each day should be marked by a few well-defined situations in which his co-operation is insisted upon. If he cannot oblige at any particular time, his participation should be terminated.

In the family, typical routine situations are the family meals, which should have a certain formal and definite character which demands that the child fall in with the habits and customs of his family. The more outspoken and articulate these customs are, the better it will be. A very young normal child, or a hyperkinetic child will, to begin with, not be able to conform to any "ceremony" or custom-determined occasion. If his behaviour becomes intolerable, he should be taken out of the room, not as a punishment, but in order that he may finish his meal on his own, either with help from his mother or in the company of one of the other members of the family. He should, however, be allowed to return to the dining-room if he so wishes and seems to be in the position to adjust to it a little better.

71

It is of the greatest importance that such measures are not handled punitively, but as a matter of simple consequence, in the attempt to allow the child to adjust gradually to those forms of behaviour upheld by the group in which he lives. As a hyperkinetic child usually has an intense desire to be with others, because of his strong force of sympathy, he will of his own accord want to rejoin the group. If unnecessary frustration is not added and his own pace of progress, possibly of necessity a very slow one, is conceded to him, he will in time be able to adapt to the demands of the family-meal situation.

In all attempts to help the restless child, it is of paramount importance that he is completely and satisfactorily reassured that he is not excluded from the love of the parents or the rest of the family. Parental love and concern must be maintained, even though a given situation may demand his temporary exclusion until such times as compliance is more readily available. It is a very different thing for a small child to be sent out of the dining-room than to be taken out by his own mother to finish his meal on her lap or in her presence in another room. We need not greatly fear, in this type of child, that he will then always want to be fed by his mother, because his wish to join the group will remain strong and persistent. Should there be any danger of an over-attachment to his mother, however, some other member of the family can accompany him in his moments of exclusion, until finally he may be willing to accept solitary exclusion while waiting to be re-admitted.

There is another way in which support for the force that underlies all intellectual, discerning and conceptual development can be fostered in the restless child. This consists of simple therapeutic exercises, such as getting the child to experience mirrored forms and reversed procedures. For example, he might walk forward with a stamping gait, possibly counting his steps. At a certain moment, he should stop and retrace his steps, while going backwards. Such exercises are capable of many variations, but all should aim at the basic experience of arresting a forward-going movement and then reversing it exactly. The same principle is present in so-called mirror-exercises in which the child walks in a semi-circle, for example, in one direction and then mirrors the same form in the other direction. It is helpful to draw these forms on the floor with a piece of chalk, so that the image of the mirror-activity may also gradually become a visual experience to the child.

Exercises of this kind can be developed further as the child grows older and blackboard and paper can be made use of. The mirror-picture has a particular strengthening quality for the development of antipathy in the sense in which we have spoken of it as the ability to hold off sense-perceptions, to develop discrimination and the ability to differentiate. The "mirror" principle can finally be applied to speech in that words and possibly short sentences are spoken normally and then backwards.

Unlimited variations of such exercises, if practised adequately, will provide considerable support for hyperkinetic children.

5. Autism

A little boy stands at the window. One hand holds on to the curtain which is swished in front of his face with rapid, flicking movements. The movement stops. Otherwise, there is no change in the boy's posture or attitude. He does not seem to have noticed that I have entered the room. I call him by name. He does not appear to hear my voice. There is no reaction. I go up to him and just as I am about to put my hand on his shoulder, he slips away. He gives a sudden piercing shriek, and with a leap, flings himself on to the bed and burrows under the bedclothes. For a while, he bounces violently on the bed, but this ceases. He lies with his face down. There is no motion, no sign of life. A little later, he slips off the bed and begins to wander about the room. I have meanwhile sat down at my desk to write. He passes my desk, and handles various objects, but of me he takes no notice. He pulls a piece of paper from beneath my left hand, and when it gets stuck in my cuff, he lifts my hand as if it were a mere object and extracts the paper.

When I turn round to him, he looks past me. His face is well-formed, his head large, his eyes big and beautiful. His features are regular and his body well made. His hands and fingers are delicate and translucent. There is something beautiful and dreamlike about his appearance. He reminds one of some pre-Raphaelite paintings.

He goes back to the window and stands motionless. He seems to have forgotten everything around him and for a time, holds his face against the window as though transfixed. Then again, he sets out on his aimless wandering about the room. He sees my camera on the windowsill, fetches a stool to climb up and takes the camera down with a swift and skilful movement. He handles the various controls deftly, winding the film on, looking through the appropriate aperture and turning the lense-mount to focus it on various objects in the room. He seems to be quite expert at handling such a complicated piece of machinery as a modern camera.

He then replaces the camera and goes from one light-switch to the next and switches the lights on and off. I swivel my chair around in his direction, but whenever I attempt to address him, there is no response at all. Yet, he now comes up to me, still avoiding my gaze, turns round in front of me and with his back to me, climbs on my lap and leans against me. When I put my arms around him, he accepts it and nestles comfortably into my hold.

I rise and get him to sit on the chair opposite and tell him that I would like to see his feet and that he should, therefore, take off his shoes and socks. He complies with astonishing deftness and speed and I am impressed how well he has understood what was wanted of him. When, after the examination, I ask him whether he can put on his socks and shoes again, he suddenly says in a clear voice, "I put on your socks".

I bend down and put on his socks. He has already arranged his two shoes neatly in parallel in front of him and when his socks are on, he lifts both feet and inserts them deftly into both shoes simultaneously. He pauses for a moment and then lets out his piercing scream, but this time he does not stop immediately but goes on screaming. I realise that he is confronted by the dilemma of having set out on a completely symmetrical procedure and he is now unable to tie both shoelaces at the same time. I then say, "I shall tie one of them for you". At which he stops screaming, bends down and, while I tie one shoelace, he watches how I do it and ties the other identically in mirror-fashion.

He returns to the window, takes hold of the curtain with one hand and swishes it in front of his face, rhythmically and rapidly. He gives no sign of awareness of my presence. Our little encounter might never have taken place. It is as though he had always been standing at the window, looking out and swishing the curtain.

Typical reports describing children like this little boy include the following observations. They avoid looking into one's face. They avoid communicating in any way. They appear not to understand or even hear what is said to them. They usually do not speak, but if they speak, speech is bizarre. They use words as though they were objects, playing with them as with marbles, pearls or precious stones. They move them and turn them round and take pleasure in odd combinations of words, but they do not use them as a means of communication.

A striking thing about their peculiarities of speech is the transposition of personal pronouns. Children like the little boy described above frequently avoid the use of "I" or "me" in referring to themselves. They may use their own name to do so, but very often they will refer to themselves as "you" and the other person as "I", e.g. "I put on your socks!" meaning, "You put on my socks!"

Signs of intense anxiety are often displayed for reasons that are not at all obvious to the casual observer. Upon closer scrutiny, it is apparent that these anxieties arise frequently in connection with obsessions in the nature of extreme fastidiousness, such as placing objects exactly in parallel, or maintaining certain spatial relationships between the articles of a room, or a particular order of things. Dependence on the maintenance of "sameness" in their surrounding is at times astonishingly marked.

Other impressive features frequently to be found in these children are the intense curiosity and interest in mechanical things, and the often extraordinary skill with which these are manipulated this, in contrast to their apparent total disinterest in people, the lack of confrontation with any other as well as with their own Selves.

The excessive avoidance of contact and communication on the part of autistic children with any other person is, however, modified to the extent that they are often open to the type of contact an infant would enjoy. Autistic children do not refuse tactile contact of a cuddling kind when there is no insistence on visual or verbal contact.

It is interesting to observe that children who have a total avoidance of language, who never speak nor show any sign of understanding of the spoken word, can reveal themselves to be accomplished singers with a large repertoire of songs. There are children who, having grown up in a musical environment, can sing complete arias or symphonies when given the first few bars.

The movement-patterns of autistic children have an extraordinary diversity. In one sense, their movements seem to be particularly graceful, co-ordinated and dexterous, and yet they are bizarre and unusual, executed not only by their hands and fingers, but also with legs and indeed by the whole body. Twisting, jumping, rolling gestures suddenly appear out of the blue and yet apparently not without a certain compulsion. There are children who must spin several times before proceeding further down the passage or along a path. Others seemingly need to touch the floor or some objects one or more times while passing by, even if the pursuit in which they are engaged at the time makes doing so particularly difficult or even wellnigh impossible. Needless to say, such obsessional movement-patterns lend to the child who has them a bizarre and odd appearance.

Many of these children have severe sleeping problems. Going to sleep is particularly difficult at any time for them, but on occasion it seems to be wellnigh impossible. The period of sleep may suffer reduction to an absolute minimum and no pattern is established at all. Some children cannot go to sleep by themselves, but require the presence of one of their parents. Some cannot go to sleep in their own beds but only in one particular chair in the lounge, and may only be carried to bed when they are fast asleep. Others can only sleep when they are in physical contact with their parents.

Equally, serious feeding problems are found with these children. Eating may stop altogether. Special means will be required to establish or re-establish more healthy eating habits. I know a little boy who at one point refused any kind of food for a number of days and who during this time began to pull out his own hair and eat it. He became emaciated and only just before tube-feeding would have become necessary, it was discovered that when a particular person brought him a plate full of food, and, without looking at him, put it under his bed then left the room immediately after, he ate the entire amount within a few minutes. The person then returned and collected the plate from underneath the bed. After this the child gradually returned to normal eating habits.

A little girl of four refused to eat under any conditions provided by the parents, but lay down on the floor opposite the family dog, assuming a position identical to it, and ate out of the dog's bowl by mouth as the dog did, without using her hands.

These are extremes, however, and more frequently, obsessional predilections for certain kinds of food are found. The choice of savoury or sour foods is characteristic and these will often be indulged in to the exclusion of other foods by these children.

It is obvious that children of this group are in some ways very different from other handicapped children. It can hardly be imagined that their situation is one of inability as they are "expert" in so many things. Their problem is rather that of being different, of being differently "put-into-the-world", something which is borne out by the results of attempted intelligence tests. While the normal, healthy though dull child will score a definite mark in any intelligence test, through answering questions up to a certain age-level and then consistently failing close to the age when the drop-out began, the picture is completely different in the case of autistic or psychotic children. A handicapped child whose intellectual retardation is not primary but caused by some developmental pathology, usually has a wider scatter of results of intelligence tests, but this scatter in most cases covers only a few years. A child of ten may answer all the questions appropriate to the age of seven or eight. In contrast, a psychotic or autistic child will often answer questions at a level far beyond his chronological age, yet at the same time fail on an infant level. These children typically produce a scatter in their scores in intelligence tests ranging over nearly the whole scope encompassed, demonstrating, in point of fact, that intelligence tests are not applicable in their case at all. The tests are in any case less applicable to the handicapped than to normal children.

But it is often possible with skilful testing to obtain results far beyond chronological age in some psychotic children, while others are obviously untestable, so that intelligence quotients ranging from below thirty to above a hundred and forty may be encountered. Now before any attempt is made to understand or interpret these bewildering phenomena, the developmental and environmental factors in the conditon of autism or childhood psychosis must be considered.

One typical history of an autistic child follows. . . .

"Henry was a first child. His father was a very fine, highly intelligent university professor and archaeologist, intensely interested in and dedicated to his work, a specialist in his own field, spending most of his time on research. In appearance, he was tall, elegant and of distinguished bearing.

"The mother was more out-going, a sensible, mature, well-balanced personality, good-looking, socially and culturally active, herself coming from a social and economic background of a high standard.

"Henry was born in the Middle East where his father was carrying out research at the time. The birth was normal and natural. Pregnancy had been a good and unencumbered time for the mother, who was happy and fully engaged in the social activities accruing from her husband's work.

"The child was beautiful at birth, seemed to thrive well and did not present any specific problems. Milestones were said to have been normal. Henry walked fairly well at the age of eleven months, made all the ordinary baby sounds, and had begun to say his first words between his fourteenth and eighteenth months.

"He was noticed to have been a little off colour after his vaccinations in early infancy and round about his fifteenth month had some febrile illness which might have been a form of dysentery endemic in the district."

However, these illnesses occupy a minor place in the report on Henry's history.

Now generally, the development of a child may have proceeded normally to all appearances up to the age of two or two and a half years. Then, sometimes in connection with the birth of a little brother or sister, the very first symptoms of a disturbance can be observed. In the case of Henry, the family moved back to Britain just about the time when the mother noticed that Henry's speech was beginning to dwindle. Where he had been speaking in definite sentences before, he now spoke less and less, was even becoming markedly silent.

It was only then that it became noticeable that he was making hardly any contact and that he seemed even to avoid his parents. He developed solitary habits, withdrew into corners and frequently showed signs of intense anxiety for no apparent reason.

Children like Henry will often relapse into losing whatever toilet-training might already have been achieved and the loss of normal eating and sleeping habits that can occur has been spoken of earlier. All this indicates that the child has obviously entered a stress-situation which results in some degree of, or even complete, regression. As has been said, the birth of a sibling is in some cases a contributory factor. But whatever may have caused the onset of the withdrawal or regression, it would seem that, roughly, the third year is the critical period in which the symptoms begin to manifest.

Another somewhat rare and extreme type of family history may be quoted.

"The father of the child in question is an engineer in a highly responsible position. The mother is very sensitive, a nervous, intense and labile person. The child is an only child and the mother reports that from the moment the child first quickened (in embryo), she knew that something was terribly wrong. Although nothing untoward happened during pregnancy, she felt an ominous foreboding that the child developing within her was going to be in some dreadful way hurt, ill or abnormal."

"Birth itself took place at full term, was normal and natural and not particularly difficult. The child seemed perfectly fit at birth and had no faults or malformations whatever. It cried immediately and presented no difficulties.

"Nevertheless, the mother felt that something was basically wrong. She felt that the child looked at her with a strange and hostile gaze. She herself expressed great repulsion to breast-feeding and the child was therefore bottle-fed.

"The mother communicated her anxieties, not only to her husband but to doctors as well, and was always reassured that the child was perfectly normal and was developing normally and that there was no ground for anxiety.

"The child's development throughout the first year did indeed seem normal. He had begun to sit up by the time he was seven months, was standing in his twelfth month, walking fairly well by the time he was fourteen months old, and was at the same time using quite a bit of baby-talk.

"Nevertheless, the mother still felt extremely apprehensive and convinced in herself that something was fundamentally wrong with the child. She took him to various specialists and Child Guidance Clinics, always receiving the same reassurance that her fears were groundless.

"Between the ages of two and a half and three years, the child began to show distinct features of autism. Speech became completely repetitive, he referred to himself as 'he' or 'you'. He withdrew from contact with other people, and responded neither to being confronted nor spoken to. The classical features of childhood autism developed rapidly and unmistakably."

While Henry, whose history was described earlier, managed in spite of his autism to establish himself to a considerable extent in an understanding and tolerant setting, and learned to live with others in a harmonious and constructive way and made contributions later through his work in various spheres, the latter child failed to make any progress at all. He remained completely stationary in his development as a person, never establishing contact to others, never finding a meaningful use of language or other forms of communication, but continuing in his own bizarre and remote reserve.

A third, equally unusual, history is the following:

"Johnny was born as the second child to a young couple, the father being employed in the railways, the mother having been very young and inexperienced when she married. While she was carrying Johnny, her first child was still in its infancy and one day when she was changing its nappies, it fell to the floor and died. The young mother sustained so profound a shock that when Johnny was born by normal delivery (which was rather difficult owing to his large head) she resolved that she would not handle him at all so that he would not be exposed to the same tragic fate that had befallen her first child.

"She, therefore, did not breastfeed Johnny, but fed him while he lay in his cot; nor did she ever lift him out of the cot to change or wash him, but managed to see to his physical needs while he was lying safely and securely in his little bed.

"Although the mother managed to nurse Johnny through his earliest infancy in this careful and cautious manner, she could not avoid becoming increasingly disturbed by the shock of the death of her first child and finally had to be admitted to hospital for treatment.

"Johnny grew into a healthy, beautiful, large-headed child but did not develop either speech or contact to other people. He would sit quietly, rocking to and fro for hours on end, and hum to himself in a dreamlike, repetitive manner. He was withdrawn to such an extent that it was thought that he might be blind, and for some time, it was considered that he was deaf. There was no reaction to sound although he hummed his unrecognisable tunes to himself."

Johnny is one of the very few autistic children I have known who has made a complete recovery from autistic withdrawal. He is now a perfectly outgoing, happy, communicative little boy, with an ever-increasing use of speech, somewhat retarded, because of a slight hydrocephalic brain damage, but giving a completely normal impression.

When an attempt is made to understand and interpret the syndrome and phenomena of autism, the question of aetiology must again be disregarded. It is likely that the symptoms of autism arise on the basis of a variety of organic as well as psychological factors. For example, in the three cursory histories given, a purely psychogenic form of autism may be discerned in the last case, a psychotic syndrome on the basis of a genetic, metabolic or other constitutional disorder in the second, and in the first case, autism may have developed as the result of a slight degree of encephalitic brain pathology, or other organic factors.

It would never be doubted, of course, that severe emotional or physical neglect can seriously injure and impair early child development. On the contrary, there is rather more reason for astonishment when injuries of such a nature do not deliver serious blows to a developing child. But there is the other question—which remains an open one —as to why an inappropriate attitude on the part of the mother, be it lack of motherly warmth or emotional instability on her part, should be the cause of autism in her child rather than of other emotional or "nervous" disturbances.

Before we can approach the core of the problem, yet one more general aspect should be discussed. For a number of years, there has been a trend to interpret autism as an inability, more particularly as a peculiar form of perceptual inability, or as disturbances in the higher mechanisms of perception which prevent the child from perceiving and grasping situations in a way necessary for normal communication between people. Such perceptual inabilities do exist on a developmental basis and can constitute considerable and serious developmental problems in childhood.

Particularly the aphasias, which are forms of auditory imperception, fall into this group. There are also other forms of perceptual inability which play between motor and visual co-ordination, and yet others which are related to the forming of concepts in connection with hearing and the experience of language.

We shall consider this later on, in a specific chapter on perceptual inability, as the connection to the phenomenon of autism is probably only causative.

There are children in whom autism may develop on the basis of an underlying aphasic condition or other conditions of disturbed or underdeveloped perceptual ability. Yet, the disturbances, hindrances or limitations in perceptual ability can of themselves hardly be regarded as autism. They rather play a similar role as blindness and deafness, or for that matter, emotional impairments or organic conditions.

Whatever the cause may be in these instances in respect of aetiology, my concern here is with the appreciation and understanding of the syndrome itself against a background of developmental pathology.

It is required to know what the basic condition of an autistic child is and how this condition can be understood in relation to his development. From this angle, certain observations may be relevant. It has often been stated that there is a typical family background where autistic children are concerned. Autism in children occurs more frequently in intellectual circles and in so-called upper social classes, but it is well known that autism is by no means exclusive to them. Autistic children have been found in all social strata and, now, practically all over the world. The question has arisen whether or not mental illness in the parents or family plays a certain part, but even if this were the case, it would be something pertaining more to the aetiology of the condition rather than to an understanding of the actual symptoms and phenomena. It is, however, fairly correct to assume that, relatively speaking, more professional people than others are parents of autistic children and it shall be seen later that the intellectual and scientific orientation of life today might have a specific bearing on the phenomenon of autism.

Another puzzling feature is the preponderance of first-and-only children in the group suffering from autism. Again it is by no means exclusive to first-borns, yet it is not without significance. An interpretation of the child's place in the family in relation to autism shall later be attempted.

Probably the most important point to consider, inasmuch as we want to regard it as a pathology of development, is that of the actual time of onset of autism, and here, to begin with, opinions differ considerably. There is the strongly held view that childhood autism is present from the moment of birth or, at least, from earliest infancy. There is a tendency to distinguish primary and secondary autism, in which primary autism is seen as an innate condition and secondary autism rather as a form of reactive behaviour. Among hundreds of histories of autistic children, I found only a small number of cases where there was a faint hue of autism already in earliest infancy. More frequently, parents seem to have noticed nothing odd in their little one. In other rare cases, the symptoms of autism only become manifest in the pre-school or school period. Between these two extremes, nearly all typical and classical forms of childhood autism appear, as it has been said, round about the second and third years of age. It is then that the most dramatic symptoms appear, and an attempt will be made to find the meaning of this through an understanding of that particular phase of child development as it normally proceeds.

In the encounter with an autistic child, the symptoms of avoidance are probably the most striking. Such a child avoids looking at one, he shuts himself off from communication by means of language or sound, and he avoids involvement in any situation. Yet, he seems so intensely sensitive, exposed and vulnerable. At the same time, he is so alert to and masterly in handling mechanical objects as well as the movements of his own body.

How can the tensions between these diversities and polarities be interpreted?

That an autistic child cannot do certain things is less impressive than the fact that he actively seeks to *avoid* doing certain things, and yet is compelled to do things that fall out of the ordinary scope of motivation.

An autistic child does not seem to be integrated into the social pale and, even more markedly, does not seem to be related to himself as a person. The failure to develop an appropriate experience of Self has time and again been described as central to the symptomatology of autism, and it is this that I believe to be specific to the characteristic phase of the visible incipience of autism—the period between the second and the third year of childhood.

It will be helpful at this point to recall some of the aspects of normal child development which were discussed in the first chapter, when the development of consciousness in early infancy was considered. We held the view that, to begin with, consciousness is diffuse, spread out, still to be centred, and that early child development entails a gradual withdrawal of consciousness from its peripheral nature, in which it includes the other person, to a state of centred consciousness which *ex*cludes all but the Self. This process culminates at the moment in which for the first time a child refers to himself as "I".

It is obvious that a child does not learn this word "I" as he learns by means of imitation all other words and designations. He hears words used to designate people, objects, actions and gradually his own original, "international" baby language converts to his mother-tongue. It was seen that the power of imitation is, so to speak, the wellspring of this next stage of development. In total contrast to this, a child never hears the pronoun "I" in reference to himself, nor is anyone else addressed by others as "I". Yet, when the experience of Self dawns, the child uses the pronoun "I" to himself at a time when his reason is by no means sufficiently developed to the point that, through the process of deduction, he could arrive at the realisation that he might call himself "I" as others so call themselves.

It is astonishing that the uniqueness of the experience of Self in early childhood is not always immediately recognised. It is the only experience that is not directly stimulated by what the senses perceive. It develops purely and immediately from a child's own inner development. However, the significance of this developmental step does not emerge if no attempt is made to see it against the background of wider or total development of a child.

So far, only early infancy has been considered. It should be noted that the first experience of Self occurs before the child is of actual nursery-class age. At that stage, a child is not open to logical approach. As we have said, his argument is rather of goodness than of truth. It will take him a number of years before he has matured sufficiently, even purely from the physical point of view, to use his intelligence and bring his reason to bear on things. The child who is not subject to too great a pressure to develop his intellect will maintain up to the fifth, sixth, and even the seventh year a mode of experience in which fantasy still plays a dominant part. When he is seen at play, it will be apparent that *meaning* has not yet succumbed to "cause" and "fact", and that he is still at liberty to ascribe alternating and ambivalent values to things by dint of his power of imagination. In fact, this is the very essence of play in infancy and childhood.

Even a pre-school child will still be inclined to invest inanimate objects with soul, because he has not completely detached his experience-of-self from his surroundings.

Although at the age of four or five, a child may no longer hit out at the table against which he has bumped his head as he might have done at the age of two, he may still readily give in to the temptation to take as actual fact or truth what he wishes to be, rather than what he has observed. This is the case not only in situations where shame and guilt are involved, but even in situations that are free of these elements, a thing which can frequently be observed when children are allowed to express themselves without constraint.

It is only after the child has entered the school-age phase that a fundamental change sets in. At this stage, reality as it can be experienced through the senses begins to impress itself on the child and dominate his experience. It is interesting that this change in a child's way-of-experience is expressed by or coincides with a change in his physical form. The proportions in the body of a pre-school child are markedly different from those of a school-age child. Whereas in the former of the two, chest and abdomen form a solid block right down to the hips, in the latter the waist slims, and with it, he begins to present the archetypal human form as it is known from classical Greek sculpture.

In this particular phase when this human form emerges, albeit still to become singled out as distinctly male or female, there is too a freeing of the mind for intellectual pursuits. The child generally would seem at this time to enter a new existential phase. Nonetheless, throughout his early school years, he is still distinctly a child. His relationships to other people are by no means those of *one* person to other persons. He is still enveloped in the "generally human". He accepts, and drinks in without question, the information and guidance that comes from the adult world as well as from his older, more experienced peers amongst children.

A child at this age still retains qualities which an adult finally and completely loses. The qualities of both sexes are maintained as one, and are not yet at variance with one another. A sense of awe, a general lack of prejudice, and the ability to give oneself up to a world from which he is yet to set himself off as the one-apart-from-all-others, are his attributes. The world and other people do not appear to the child at that age in terms of truth, but rather in terms of *aesthetic* values. The question, not necessarily so articulated, is: "What can be admired? What is the desirable and heroic image on which I might model myself?" The internal landscape, as it were, through which a child of early school age passes is one in which values are derived from his sense of the beautiful, the desirable and the aesthetic.

Not until the middle of the school period does a child gradually grow into an adolescent world, beginning around his twelfth or thirteenth year. Here, the separation into the sexes becomes not only physically and biologically manifest, but it takes place in the psyche as well. A child is from birth either male or female, but, the experience of this has none of the stark significance it has for an adult. The qualitative relationship between the male and the female elements begins to unfold only at puberty.

The truly fundamental change, however, most characteristic of this period, occurs in the child's relationship to himself and to other persons. He becomes lonely and isolated. The ability he had in his younger years to identify with what he imitates largely recedes. He becomes concerned with himself, with his own identity.

Between the fifth and thirteenth years, a child lives, so to speak, by virtue of abilities such as the power of memory, or of imitation, of learning, or of absorbing. A young adolescent begins to see himself determined by the state of his emotions. It is his emotional reaction to things which now matters most. In this new mode of experience, he is yet to establish what an adult person knows as the "core" of his existence as a person, his innermost self. A child's sympathies and antipathies are the central and moulding agents in the time between puberty and early adolescence. Accompanied by sexual development, this phase is also marked by the loss of harmony in the human form. Limb growth accelerates excessively in alternating phases, as does that of the physiognomical features, generating characteristic disproportions in the harmony of the body, until a fresh harmony is achieved when a youngster has reached the age of eighteen or twenty.

During this period the realisation or experience of *self* (in the sense of ego) commences to pervade. Although personality continues to develop and only begins to unfold later on, in many respects, with this, the point has been reached which is unmistakably the conclusion of adolescence.

On the one hand, it can be sensed how the experience of one's own and central egohood belongs to that phase of maturity between the eighteenth and twenty-first years and is indeed its core and ruling characteristic. On the other hand, there is that singular moment in a young child's life between his second and third years, which can be witnessed in each young child anew, when his sudden awakening occurs to the experience of himself as "I".

Initially, it is not easy for an adult to place himself in the situation of this sudden awakening of the experience of self in the young child, because he has usually forgotten the moment it occurred in his own life. But it is feasible through the use of imagination, to reach tentatively towards a child's experience in this crucial moment by briefly reiterating the general situation of a child between two and three years of age.

He has learned to stand and to walk. He can move about and name the objects in his small world. Although he has moved away from the early phase of all-embracing consciousness of omnipotence, which is his in spite of his helplessness, the world around him is still "*he*". He is still able to make objects animate and to change them into whatever he fancies and to charge things like dolls, bits of wood, tables or chairs, with higher significance. In short, his is still a magic world. The two-year-old is Master and Lord in this world. It is his Garden of Eden. His total dependence is at the same time his over-lordship over the world which is his.

Into this harmonious and absolute state, breaks the sudden precipitation of the experience of "I", "I am". His eyes are opened, to speak with Genesis, and the child has an experience of his centred Self as being the one who experiences, separate, different from the world around him, which is all at once no longer *he* but the *Other*.

It is vastly important to note, however, that in spite of this dramatic and archetypal moment, child development at this stage is still far from having achieved any degree of centredness of either experience or consciousness otherwise. This can only become established in the course of many years of development and maturition.

In fact, it becomes apparent that the experience of one's own ego between the second and third years of childhood is an impact of almost grotesque difficulty, something that can in no way as yet be mastered. The question here is not why things sometimes go wrong and lead to autism. Far rather, it is how can things possibly go right? How can a two to three-year-old child ever cope with the overwhelming impact of his own ego for which he is so unprepared and unequipped?

Perhaps a brief word of explanation ought to be inserted here of a probably unusual interpretation of the story of the Fall of Man. Contrary to the actual wording of the text, the great Myth at the beginning of Genesis has been popularly misinterpreted as describing the guilt of man in terms of a sexual "fall" which necessitated his expulsion from Paradise. However, the Myth says that Adam's expulsion from Eden was brought about as a result of his eating the fruit from the Tree of Knowledge. His eyes were opened and he knew his nakedness. Thus at too premature a stage in his existence, he was aware of his own potential divinity, hence, the Lord God says: Behold the man has become like one of us, to know good and evil.

That the Fall of Man has been interpreted popularly as having taken place by reason of sexuality can be understood in relation to Freud's discovery of infantile sexuality, which does not imply sexual activity but rather the knowing of creative power, that power which the child experiences as his own ego, while he is still the creator, the maker of things rather than knowing himself to be a creature.

The story of the Fall of Man possibly indicates that in terms of the evolution of mankind, this creative power was injected into Man prematurely, an assumption supported by the myth of Prometheus. In terms of the individual, this is the existential problem that confronts every child when he goes through the developmental stage between his second and third year—the fact that something happens to him at so tender an age with which he is only equipped to deal eighteen to twenty years later.

It may be possible to experience in the child who becomes autistic, a panic reaction to the overpowering and precipitate awakening of the own ego-experience.

Normally, a child is sustained while undergoing this dramatic experience by the relationships between himself and his mother, and his family and himself. Instinctively, mothers are often more or less subconsciously aware of the individuality, or of the ego of their child even before it is born. In some mothers, this awareness can precede conception. In others it coincides with the moment of conception, and more frequently evolves during pregnancy.

The instinctive awareness of mothers of the nature of the coming child as an individuality is rarely an articulated or defined one, nor can it so readily be put into words, yet it plays a decisive part in the degree of protective relationship the mother affords to her child when his "moment-of-truth" comes with the first ego-experience.

On the one hand, the relationship of a mother to her child is of the nature that I have just described—she has some instinctive knowledge of him as an individuality. But on the other hand, the relationship of the mother to her child is equally determined by his helplessness, his physical requirements, and his extreme dependence. It is this ambivalent relationship which will hold a child at the crucial moment, in the experience of humility, of necessities-of-life, of the fact that he is, after all, a creature of this earth, and which will at the same time support his experience of ego and link it to his physical and biological development.

(The first two thousand years of Christian history witnessed the attempt to balance the overpowering event of the Divinity having entered into human existence through Jesus Christ, by intense exercise in devotion and humility, so that men should become capable of

89

accepting and facing the realisation that God is no longer in the heavens, but in human existence itself.)

In other words, in the mother-child relationship, the balance is held for the child between the experience of the core of his existence—his Self which may be said to be of "divine" nature—and the state of being a humble, dependent creature of this earth.

When it is seen from this aspect, it may become clearer why childhood autism occurs more frequently when the family, and more particularly the mother, are rather more intellectually or scientifically minded. The pressure of a modern scientific outlook may be such that a mother, in her intellectual honesty, cannot confess to having any instinctive or intuitive feelings about her child's early individuality. This leads either to a weakening of instinctive, intuitive faculties or an over-intense intellectual and conscious effort to suppress these elements in mother-love. As a result, that part of the mother's love for her child which is called forth by his helplessness and dependence upon her is objectified. A mother regards her child's needs and dependence upon her as factors which have to be dealt with effectively and efficiently, not involving her emotionally to any great extent.

Such attitudes which have, for a time, been fostered by science may lead to a weakening or even a breakdown in the sheath of love which is so essential for a child in order that he may pass through the phase of his first ego-realisation successfully.

In reading the autobiographies of single or firstborn children, the outstanding and excessive burden of the responsibility of ego-experience heightened and intensified, is encountered, because the firstborn is the one who is to take the place of the father in due course. This means not only the succession to the physical, biological father, but it would seem to imply to Fatherhood as such. It is obvious, therefore, that firstborn or only children are more prone to autism than second or later children.

In a similar manner we can interpret the fact that so many autistic children are particularly beautiful with large beautiful heads, and are potentially intelligent and gifted. As lightning is more likely to strike the highest steeple, so is the overpowering force of ego-awakening likely to be of greater impact to just these children, and hence there is increased danger of autistic development in this group.

I have written an inordinate number of words on childhood autism. This is probably a sign that it represents a special challenge, so that, in a sense, one feels more involved in, more responsible for this condition than for many other conditions. It is probably also a sign of our lack of detailed knowledge of autism in the child.

To sum up, the attempt has been made to interpret childhood autism as a panic-reaction to the moment when the ego first makes itself known to a child between his second and third years. If circumstances are unfavourable, either because of a specific family situation or because the child is constitutionally vulnerable, or as the after-effect of an illness, the ego-experience is not only merely an intense and dramatic one, but an overwhelming and threatening one too.

In consequence of the panic-reaction, there can develop avoidance of the realisation of the self. There can be a self-denial, an increasing resistance to the integration of the ego, to the extent that, for example, the child uses the pronoun "I" impartially, as though it were a designation like any other, for any other person, or refers to himself as "you" or simply by his own name as though he were someone else. This transposition of the personal pronouns is perhaps the most unique and classical demonstration of the panic-reaction against the dawn of one's ego-experience. The failure to lodge the ego-experience at centre is probably the core of childhood autism.

This understood, the avoidance of inter-personal relationships becomes itself understandable, because we can only enter into a relationship with other people to the extent to which we realise or experience ourselves as a person. Further, the intense preoccupation with the world of things and technology may be seen as an escape or flight, because these things do in no way remind the child of his own ego-nature. In fact, in the inanimate world he can display, practise and develop his faculties, skills, manual dexterity and intelligence freely, without coming up against his threatening developmental problem, that of his encroaching self-awareness.

We can also learn to interpret what appears to be highly bizarre motivation, either in their use of language or of movement, as a manifestation of an incipient integration of intellectual and motor-development without that conscious guidance that stems from the ego or self. Things are done partly out of a satisfaction deriving from performance of functions and the use of potentials and partly simply because they can be done at all, since motivational development is severely impaired.

The somewhat happier side of the autistic child in which he indulges in his functional pleasures and in the satisfaction of pure performance is, unfortunately, limited, and anxiety, severe despair and profound unhappiness, for no tangible reason, are generally preponderant. But these things are better understood if attention is focused on the *panic* roused by the confrontation with the self which is, metaphorically speaking, precisely that which leads to expulsion from Paradise.

The so frequent phase of regression that seems to herald overt autism is obviously a reaction, a flight back into the security of an earlier phase when ego-integration does not yet loom and threaten. The strange and striking dependence on and desire for sameness in autistic children seems to be a ritualistic attempt to relegate all human existence to purely mechanical or geometrical forms, to a world of things rather than to a world of *being*. Avoidance or negation can so dominate the existence of an autistic child that it becomes the over-ruling quality.

However, there is another element in a child who suffers from autism. It is ambivalence, because it is human nature to want to maintain ambivalence under nearly all circumstances. Therefore, because of the largely suppressed, yet inherent longing for normality, an autistic child is forced to seek refuge in a host of taboos and rituals. Much of the performances of autistic children becomes understandable when this ambivalence is realised. His bizarre modes of behaviour or use of language are attempts to achieve some kind of communication which does not bind or involve him. His is a Janus-faced situation in which his inherent wish for normalisation battles with his panic in face of the demands entailed by ego-integration and its avoidance under all circumstances. The conflict and the tortured condition of the child suffering from autism are only too painfully apparent.

From all the foregoing, there emerges one definite, direct, specific and very simple therapeutic approach to the autistic child. Parents and teachers often adopt this approach intuitively, but it can be developed more specifically and effectively when the above interpretation of childhood autism is fully understood. This therapeutic attitude entails never confronting the child directly. We should never attempt to look into his eyes and address him as we would another person. Rather it is necessary to learn to see that the autistic child is not truly "in himself", and that we can reach him if we address ourselves to his peripheral self, to that which is not centred. Thus, when we address the child, we do so while looking the other way, or if we want him to come to us, we look in the direction we would have him take. Speech should be gentle, non-committal and vague rather than forceful and to the point.

When approached thus, the autistic child is noticeably relieved and can more easily co-operate and fall in with necessary demands. I have seen very slightly autistic children becoming completely with-drawn as the result of a direct, forceful, though warm approach— such as a person taking the child by both hands, looking sternly into his eyes and admonishing him to co-operate and to make a positive effort. This kind of treatment may be called for under certain other conditions when the child has gone astray in his development and has to be set upon the right path, but for the child suffering from autism, it is totally inadequate and can have disastrous results. Indeed, I should like to reiterate—it is the one approach that must under all circum-stances be avoided with autistic children..

(This creates a specific and difficult problem when childhood autism is combined with deafness, because the deaf child requires a face-to-face approach if he is to be helped in the sphere of speech and language, but it is, nonetheless, so detrimental where there is autism that it often precludes the deaf psychotic or autistic child from success-ful speech therapy and this further aggravates the problem of non-communication.)

The indirect approach described, if persistently extended to the child by all who constitute his environment, will allow him to make use of whatever positive wish for contact he may harbour and can in itself, lead to an improvement of the child's developmental situation.

There are other attitudes which are not so specific and are less easily maintained towards the autistic child, although once there is some understanding of the nature of childhood autism, the value attached to these attitudes may increase. I am referring to something I have mentioned previously, i.e. the necessity for those who con-stitute the child's environment to maintain those forms and standards of conduct and behaviour they find appropriate to their own needs, even if the child seems to react badly. With the most stringent honesty, must one distinguish between those things one feels one has to main-tain for one's own sake and those one maintains for the child's sake. This means that there must be a clear distinction between what one does for the child's good and the limits of one's own capacity for toleration and endurance.

It is of no help to try to wean an autistic child from his obesssions or fixations, which are common symptoms in childhood autism or psychosis, expressing themselves in a variety of ways. As long as the obsession or fixation is neither dangerous nor harmful to the child as well as not entirely unbearable to those he lives with, there is no virtue in attempting to stamp it out. Weaning from one obsession, if successful, will in all probability create another which could be more harmful and dangerous than the first.

On the other hand, if a fixation is not harmful to the child, but so taxes the capacity of endurance of those he lives with that they become increasingly irritable and disturbed, it presents a reason, provided it is fully and plainly admitted to oneself, to try to restrain the child and adopt means to wean him of the obsession.

On the other hand, once the nature of childhood autism is understood, the capacity of endurance will be the greater and we shall be able to bear with the child's negative and problematic modes of behaviour with more ease and thus adapt to the child's peculiar condition.

The first task that confronts both parents and teachers is to learn to live with the autistic child in a positive and sympathetic way, without undoing the constructive environment which a normal family or normal school routine presents to a developing child. It is not helpful to an autistic child if his family gives in to his every whim and adapts its life to his way of life, for when the family does so, a truly "crazy" environment is created and the beneficial influence of a "normal" environment is foregone. Because of the need for sameness, the child will react badly to any change in his environment, and will thereby perpetuate an undesirable situation.

However, those changes or modifications in the way of life that parents feel they must make for their own sakes will be more easily accepted by the autistic child if they are convinced of their necessity.

In every single case, both family and teachers will have to discover where to give in to the needs of the child and where to uphold certain forms of structured environment. The ultimate decision must be based on an attitude of warmth and understanding and on the degree of empathy attained to with the particular child.

The establishing of an acceptable mode of living-together will depend on the ability of those concerned to allow the child to experience the security of love in spite of the difficulties he has in conforming, in spite of his odd behaviour, which may often have to be curtailed, either for his own sake or for that of his environment.

This "agreement", based on warmth and empathy, can be developed and fostered when we have learned to experience the situation of an autistic child as a fundamental problem of individual child development, but equally as a problem of the development of man as such and one in which everyone shares.

This realisation will call forth in ourselves the peculiar intensity of compassion that is needful if we are to live with the autistic child especially. When a relationship of security has thus been built up around the autistic child, more direct therapeutic efforts can be attempted.

As the lack of inter-personal relationship and the avoidance of contact with the other constitute undoubtedly one of the most serious and painful symptoms of autism, we do well to make a specific effort in this respect. While adhering strictly to what has been outlined as avoidance of direct confrontation, we can, for instance, by the use of simple games, exercise reciprocal action between child and therapist.

Depending naturally upon the age of the child, a variety of games can be used to establish a to-and-fro relationship. These can be ball or ring games or simple dances connected with songs and rhymes. For children not ready to enter into a game-situation, however simple, clapping exercises between child and therapist, crosswise and parallel, can be helpful or even more rudimentary measures, such as infant games with fingers and toes, can be employed.

In these early and basic game-situations, it is of help to let a child sit on the lap of the one who does the exercises with him, with his back against the therapist's chest. This is a position of contact even the most severely autistic child will be inclined to accept or even spontaneously seek.

Games between two people, ball games, ring games and the like can later be led over into team games and ordinary sport activities and finally to such games as tennis as well as fencing, of which the latter has proved to be markedly helpful for somewhat older autistic children.

There is much variety here and the more individual initiative and inventiveness there is on the part of the therapist, the better the chances of success.

Music therapy is of particular value in the treatment of autistic children. Very severely autistic children who completely refuse to make use of language are often very musical and able to sing. Such children may sometimes be coaxed into a first kind of conversation on the basis of the duet. The therapist hums the first bars of a melody which the child may continue and thus, without looking at one another they can sometimes establish a to-and-fro in a purely musical realm. With some of the most severely disturbed children, similar initial contact can be made by tapping a rhythm with the finger on an object, table-top, cupboard or wall. Occasionally such tappings will in the end be answered in one way or another and they will provide the starting-point for rudimentary communication.

Music therapy, however, has more direct possibilities too, and to guide the child's musical experience from the seventh down over the tonal intervals to the fourth and third can be of great therapeutic value when done expertly.

Many other more sophisticated therapies can be attempted under suitable and expert conditions, including coloured light therapies and most especially, curative eurythmy.

It is equally important to allow autistic children fields of self-realisation and self-expression which may exceed their ability to communicate directly with other people, which is the case when these children are gifted artistically in drama, painting or drawing and the like. They should be given every help to establish themselves in the field where their particular gift lies because once they have achieved a degree of certainty and skill of expression, another door is opened to therapeutic guidance.

Very many autistic children have islands of outstanding intellectual performance, and it is, of course, helpful for them to realise their potential. However, we must not be misled to the extent of thinking that the efficient use of intelligence is in itself equivalent to healing and recovery in childhood autism. The essential question is not that of the child's intellectual or scholastic development but of his emotional and contactual development. This will continue to require adequate en-vironmental guidance so that all of a child's potential of personality can be fostered and encouraged in every possible way.

Before going on, I should like to mention one thing which has a unique power of drawing autistic children out of their isolation. This is the puppet-theatre. The most severely withdrawn child will become indistinguishable from other normal healthy children in his reaction to a puppet performance, for he can enjoy and participate in a great variety of dramatic human situations without committing himself. It is not going too far to say that this acts like balsam to his own torturous situation. The release which a child experiences in the puppet-theatre may be a passing one on the surface, but nonetheless, his moment of complete surrender to what is enacted on the puppet-stage is of profound value.

The best possible environmental benefit is probably afforded to an autistic child if he can mix with children suffering from other types of handicap. There is a tendency to establish special units and schools for autistic and psychotic children. Here, however, the autistic child experiences just the duplication and exaggeration of his own problem. But when he can live and go to school with children suffering from entirely different handicaps, he can derive specific and often astounding benefit to his own development. This is particularly the case if autistic children and mongol children are mixed. The mongol child is inclined to be loving, outgoing, full of mischief and extraordinarily rich in contact, and particularly in his favour, he is not put off by lack of response from his partner. This makes him the ideal companion for an autistic child, whose lack of response can never dampen a mongol child's enthusiasm. On the other hand, when an autistic child is expected to participate in ring games, for instance, he will be inclined to stand aloof and not want to mix. If, however, he sees his mongol companion participating enthusiastically but because of his clumsiness, not managing to adhere to the prescribed patterns, his own obsessive need of order and form may compel him to join the circle, purely for the sake of making the mongol child fit into the patterns more effectively.

We have seen numbers of autistic children learning to participate actively and with interest in situations like these to the mutual benefit of both sides which is an initial but infinitely important step in their social adjustment. This principle of non-segregation can be extended in a variety of ways all of which will be helpful. There are children, for instance, who, having suffered from hypercalcaemia in early infancy, develop highly differentiated skilful defensive speech. They will be inclined to be most articulate and fluent in any situation in which their anxiety is aroused. Thus, in their confrontation with autistic children, they will maintain a flow of one-sided conversation, and provide an opportunity for the autistic child to experience language which otherwise he often forgoes, because he cannot bear to be "talked at". Those who live with these autistic children who have withdrawn into silence often become silent themselves because of the lack of response, but just in this area, the hypercalcaemic type of child mentioned above is undeterred.

The encounter with entirely dependent and helpless children such as cerebral palsied children will often draw autistic children out of their isolation to a surprising degree, as the helpless, palsied child arouses active compassion in the other. Autistic children can be seen extending help and care to a palsied child which they would never do to others.

Needless to say, it requires a positive and harmoniously structured environment to bring the encounter between children suffering from different types of handicap to fruition to the benefit of all concerned, but this is to be striven for where the autistic child is committed to residential schooling. Other children with slighter degrees of autism may manage to go to ordinary schools with other normal healthy children, but here it is of the utmost importance that they are met with sufficient tolerance and that the teaching staff does not think that lack of advancement in learning can be helped by extra teaching and coaching. It is essential that it is realised that the problem of an autistic child is not one of making sufficient scholastic progress but fundamentally one of his growing up into some form of mature adulthood which will allow him to live and work with other people.

The degree of success with which an autistic child will be able to live and work with other people when he is an adult does not depend on his intelligence or abilities, nor even necessarily on his ability to communicate. Very severely disturbed, retarded and non-communicating autistic children have been helped to learn to live and work, albeit on their own terms, in sheltered and tolerant communities to the benefit of all in the community.

For this reason, it is important to re-adapt our therapeutic attitude when the child reaches the region of the fourteenth–sixteenth years. At this stage concentration on widening the scope of the child's experience in just those directions where he is particularly limited is no longer called for. He should rather be helped to make use of his fixations and obsessions in a creative fashion. Very often, severely autistic children who have grown up in a congenial and therapeutic environment will, when they are between the ages of fourteen and sixteen, take to one or the other manual craft and display unusual gifts and abilities. Weaving, pottery, woodwork, embroidery, sewing, even glass-engraving and other highly skilled crafts can be mastered by these children or adolescents to an astounding degree. They will frequently execute complicated craft procedures after having observed them for half an hour or so, without having had them explained or specifically taught to them.

Such abilities enable these young people to fill a personal place in tolerant communities and to make a contribution through their skilful and productive work.

Under such circumstances, they will also find their own social standing and become valuable members of the social community, possibly still not communicating freely or perhaps not at all, yet upholding the order of things in a fine and unobtrusive way. Thus they establish a working equilibrium in their limited and restricted relationships and involvements with other people.

6. The Blind Child

Blindness is one of the oldest afflictions known to man. Here, however, it is wished to consider blindness inasmuch as it presents a developmental handicap. This section will therefore be mainly concerned with so-called congenital blindness, that is, blindness or severely impaired sight which is present from birth. Although blindness acquired later on can cause developmental problems, the nature and degree of the problem will depend largely upon the particular period in which blindness sets in. The same holds good for deafness when it sets in at a somewhat advanced phase of a child's development.

In these situations, a child has already experienced the worlds of sight and hearing, and his picture of the world will have established itself to an extent according to the length of time he has been able to see and/or to hear. When blindness afflicts a child before the first distinct experience of ego has dawned upon him, its effects may be similar to congenital blindness. The later in time that blindness sets in, the less severe developmental disturbance it will cause.

To begin with, ideas concerning varying degrees and different types of blindness based on a physical impairment should be established. Partial-sightedness is spoken of as a relative impairment of visual acuity with which colour and form can be experienced but finer details cannot be distinguished. Colour-blindness refers to a condition in which, though there is normal sight otherwise, some, or all, colours cannot be seen or distinguished. Blindness is usually spoken of when sight is so impaired that objects and people are no longer recognisable, even though colour-and-form-perception are not entirely lost, tapering down to the mere perception of the difference between light and dark and finally to complete insensitivity to light.

Types of blindness can be conveniently differentiated into three groups, according to the anatomy of the eye and the anatomical structure pertaining to the sense of sight. The first type is based on an impairment of the geometrico-optical part of the eye, that is, the cornea, lens and virtreous-body. Either the refractive qualities of these organs can be affected, or more frequently and more seriously, the quality of transparency is reduced. Cataract or opacity in the lens, or densifications in the vitreous-body which lies behind the lens cause a scatter of light, making the perception of form impossible or at best, extremely difficult. In more serious cases, the entry of light is impeded to such an extent that the perception of colour and even light itself become obliterated. However, opacities or near-opacities can be so localised that impairment of sight is not generalised but a restricted, localised vision is possible. Under such circumstances, a child must look out of a "corner" of his eye, and he holds his head obliquely, so that his eye is in such a position as to allow the use of the least affected part of its optic system. His visual performance will be better when ambient light is not too bright, and extreme scatter of light from the more affected parts which could blur the potential vision of the less affected part, is prevented.

The second type of blindness is related to the light and colour-sensitive organs in the eye, which are the retina and the optic nerve. Here, too, the affliction can be more or less widespread and involve only part of the area of the retina and optic nerve, or its entirety. Partial affliction will reduce in most cases the extent of the field of vision, but even when the affliction covers the entire area of the retina and the optic nerve, impairment is often not absolute and remnants of actual or potential sight may persist. These things will be described in greater detail in the approach to the question of possible therapy and re-education.

The third, somewhat more rare, form of blindness is the so-called cortical blindness which results from damage to the part of the brain which is linked, not to sight-perception, but to our actual *experience* of sight. A person with this form of blindness can see everything but is not able to recognise what he sees. He will, therefore, not run into an object which stands in his way, and he will be able to pick up a minute object with his fingers even in dim light, but his recognition of the object, of its colour and particularly of its form is difficult or even impossible. Cortical blindness is a perceptual rather than a sensory handicap and can, in a way, be likened to the aphasias which will be described in Chapter VIII.

The developmental situation of the congenitally blind child is complicated by the fact that "our eyes are opened", which means, that to the normal person the world is visible and spatial to the extent that even time becomes a conscious experience with the help of visible space. This visual world in which we live and by which we know much, is not accessible to a blind child. If our world is one of space, his is one of time. Whereas our world appears to us predominantly in forms and colours, his is experienced predominantly in sounds and tones, intervals and rhythms.

The essentially and profoundly blind child is often disinclined to use his hands and fingers for tactile experience, for the exploration of his surroundings. He will rather give himself up to sitting somewhere, gently and quietly rocking to and fro, possibly humming to himself. This reluctance to reach out and orientate himself cannot be explained only as anxiety and a fear of hurting himself. It is rather due to the fact that he simply has no experience of the world as a spatial one.

His relation to the world is one in which he waits for what is to come and remembers what has been. When he sits with his head bent forward with the peculiar attentiveness characteristic of blind children, we are reminded of the attitude of meditation, but in fact, he is perceiving the flow of Time. He is, so to speak, listening to the flow of Time.

The awareness of the fact that he perceives Time as we perceive Space forms the foundation of our ability to live with profoundly blind children and help them to make some contact with the spatial and visible world. They need stimulation in their own world of time to help them to differentiated and articulate experiences. Here the most obvious and by far the most suitable medium is music. The blind child's ability in this field surpasses that of the ordinary person and his educators should feel it as an obligation to open up the world of music to him because it is both one in which he can feel entirely at home and at ease as well as one he can share in with those who can see. The same applies to literature, history and the human sciences in which he can feel equally at home and at ease.

It is obvious that the profoundly blind child can only become fully integrated when he learns to experience the spatial world as well as the audible one, the world of time. Therefore, a tremendous effort must be made to introduce him to spatial experiences and concepts. Even the most rudimentary of activities such as dressing and washing oneself are dependent on a considerable amount of spatial skill and orientation.

Efforts to help the blind child will be more successful if, with an understanding of his world of experience, on the part of his therapists and teachers, he is allowed to derive his spatial concepts gradually from his temporal ones. For him, what is *first* or *later* is more primary and salient than what is *up* or *down*, *left* or *right*, and he tends to persist in this linear mode of experience, which is the way in which we normally depict Time to ourselves. We will have to find his way from the linear and one-dimensional to the two-dimensional, and finally to the three-dimensional experience of Space. It is helpful to realise that there is not only a *visual* experience of three-dimensional space but also one based on the sense of movement and the body-image; the vertical dimension goes from the head to the feet, the horizontal from the left to the right through the chest, and the sagittal from back to front.

The practising of spatial experience in relation to his own body is of greatest developmental help to the blind child. Likewise the development and differentiation of his sense of touch is essential, bearing in mind that he needs to use his fingers more as eyes than as tools. He will certainly have to be helped to achieve manual skills, but his finger-tips must develop and retain the greatest possible sensitivity.

If the blind child is to find his way about, he should not be imme-diately exposed to stretches of empty space and expected to master them, but rather let him begin in his own room where the furniture and other objects are permanently placed and fairly close together. He will, however, also require a free and empty space where he should be encouraged to play and where he will not knock against objects and furniture at the slightest movement. This is particularly important for a child originally able to see and who therefore has retained some feeling of and orientation in space, but who is now disorientated and insecure because of the loss of vision.

For the profoundly and congenitally blind child, it is important to help him to develop his "own space" first of all, beginning with his body alone and then his body in his most intimate and familiar surroundings. It is not always easy to carry out such measures, par-ticularly when the child is in institutional care, but nonetheless, it is a primary requirement.

The tactile sensitivity of the blind child can be nurtured in a variety of ways and here, plastic sculptural activities are of considerable value. Likewise the use of the many teaching-aids which have been worked out and are available for blind children is beneficial.

The treatment of blind children by means of *visual* exercises, however, has been comparatively little explored. While it has become a generally accepted fact that even the severely *deaf* child benefits from continued and intensive hearing-stimulation and can often be retrained to achieve a useful degree of hearing, parallel attempts have not been applied in the treatment of blind children. This is all the more inexplicable in view of the fact that it is well known that the objective degree of blindness, the actual impairment to sight, does not always exactly correspond with the degree of useful sight retained by a child. There are children with objectively very severe sight impairments who have a considerably higher degree of useful sight than others whose sight impairment is objectively less severe.

This would seem to indicate that sight is not only *seeing*, but also *looking*. Sight is, of course, not only an optic and photo-chemical process like the taking of pictures in a camera, but it entails an involved psychological process as well. The nineteenth century idea that, as a first stage colours and forms are seen has long been discarded. It is known now that people and things are seen in certain shapes and in colours. That which is made conscious in the act of seeing is not, primarily, colours or shapes but familiar as well as unknown *objects* and *people* always recognised however as objects or people. From the development of painting since the beginning of this century, it is clear that our process of vision requires considerable artistic training to perceive primarily colours and shapes divorced from conceptual meaning. Cezanne's work is probably the first realisation of this fact, and the subsequent development of modern art is to some extent based on the re-education of the sense of sight.

In the attempt to become conscious of actual visual experience, it must be realised that the basis of ordinary seeing is the directed activity of *looking* with distinctly outgoing motor-qualities. When a person learns to withdraw this activity from his seeing, he arrives at a primary experience of colour and form, the one vivid and intense, the other original and striking.

For most people, this exercise takes some time to be fully mastered because into the attempt to withdraw *looking* from *seeing* enters an element of *gazing*. In gazing, something of an emotional and personal quality shines out through the eyes, and therefore, the gaze can be so telling, so meaningful, and experienced as intense communication.

When we look into another person's eyes, we do not study his iris or some other part of his eye, but do so as an expression of our being related to him in some way, as an expression of a certain surrender.

This very quality of gaze has to be eliminated as well if what is called pure seeing in artistic terms is to be arrived at.

Sight has perhaps three aspects which may be described in the following way—first of all, the world outside streams in as colours and forms which have their own impact. To ward off this impact, the activity of looking is pitted against it, and the optical phenomena are translated into experience whereby things and people are seen in colours and shapes. This is the second aspect or component of sight. In the third, harmony is established—the total original impact of what was seen has been warded off by the activity of looking and now something of that experience can be given back in the gaze.

The first component of sight depends largely on the optical perfection of the transparent parts of the eye. Opacities or refractory errors are not easily corrected by original activity on the part of the one seeing although to some extent, correction is possible. If, however, impairments of the optical parts of the eye are so severe that the result is blindness, it becomes very difficult to achieve improvement even by guided and intensified stimulation, because just this will cause the above-mentioned dispersal of light and thus an increased blinding effect. When the impairment, or lack of transparency, is nearly total, intensification of stimulus will also not be of any help. Various means of re-training sight should, however, always be attempted whatever the circumstances, because occasionally, worthwhile improvement can be obtained.

This is very different, however, in the second type of blindness which is connected to "looking" and to impairments of the nerval and light sensitive part of the eye.

Although this type of blindness often appears to cause severe and profound blindness, it is hardly ever total. There seems to be a certain parallel here to nerve-deafness, and carefully handled light- and colour-stimulation can often bring about considerable improvement in effective sight. Under medical supervision, lights in complimentary colours can be shone into the eyes of a blind child in regulated rhythm. As a result of this kind of treatment, very profoundly blind children have learned to control their nystagmus, which is the involuntary oscillating movement of the eye-balls, and to focus on the light. They have been known to learn to focus on strong coloured lights shone on to a white background in a darkened room, and also to follow moving lights, which is a particularly helpful exercise in stimulating the activity of looking which is, as has been seen, in point of fact a motor-activity. With continued treatment of this kind, quite a number of children attain to a first degree of colour differentiation and even if they do not develop so called useful sight, they have an experience of *seeing*.

This in itself is of very great value because it opens a window into the experience of the visible world. After this, it remains to be seen whether the individual child can be led further and develop some degree of useful sight, however limited.

A child suffering from cortical blindness requires an entirely different therapeutic approach. He sees everything but does not know what it is that he sees and is also unable to use his eyes to look at another person. The experience of *gaze* is not open to him because the colours and shapes streaming in from outside cannot be divested of the force of their impact, cannot be disarmed, so to speak, as they attack. It requires prolonged, careful and patient education to lead such children, first to the recognition of colours and then to the recognition of the very simplest, most distinct and primary shapes such as circles, triangles, elongated rectangles, squares and so on. Very gradually they gain their first experience of looking at and seeing objects and finally, possibly even persons. If this can be brought about, gaze can also be achieved.

Most partially-sighted and also many profoundly blind children adjust and manage well. There are some who, particularly if suffering from additional handicaps, are inclined to panic reactions resembling those of childhood autism or psychosis. These children present the symptoms of autistic withdrawal as we have described them in the previous chapter, including extreme anxiety and severe sleeping and eating difficulties. They do, however, respond well to understanding handling and guidance, and in most cases the autistic or psychotic features are overcome in a relatively short time.

Probably the most severe and most debilitating combination of handicaps is that of blindness and deafness, a condition often caused by rubella in the mother in the early months of pregnancy. Children suffering from profound blindness as well as deafness present a formidable problem. To begin with, it must be realised that it is a problem of our own communication. We are helpless in the face of such a child as our repertoire of communication is so utterly insufficient and so hopelessly frustrated.

The human adult world is over-determined by the visual and to a lesser extent, by the audible upon which, of course, language as the main means of communication is based. Therefore we feel that we are totally cut off from any possibility of contact with the deaf/blind child.

If, however, the earlier interpretations of infantile consciousness are recollected, showing it to be expanded and not yet centred, a notion of a possible way of contact with such severely hurt and handicapped children suggests itself.

While they are yet very young, they participate in the experiences of their surroundings and a mutual bond of understanding and communication can be established between such a child and his mother to begin with, which can become the basis for a further development of communication.

Under no circumstances must the use of the spoken word cease in confrontation with these children, partly because of the hope that continued stimulation in the field of language may arouse some minimal degree of hearing-development, but even more because our own way of presenting ourselves and communicating about ourselves is so completely bound up with language. We could not with any integrity enter a child's vaguely extended consciousness if we were to present ourselves as mute.

Naturally, any vocal communication will have to be accompanied not only by visible but also by tactile gestures. A child will begin to sense when we are near him or when we draw away from him and a basic differentiation of relationship in a spatial sense becomes an experience to him. Not only the sense of touch but the sense of warmth too must play a part in this interaction, because it is important for such a child that the first attempts in education are directed to the relationship with another person—the therapist in this case—rather than to the mere manipulation of objects. Human warmth is essential for communication with such "imprisoned" children.

A relationship to objects must follow such a primary relationship to another person in the course of education, and the world of things and their use and meaning must be opened up, together with the teaching of many basic skills and practical activities a child has to master. This requires infinitely patient work in repeated sessions over long periods of time, but it must be kept in mind that the learning of skills and practical manipulations can never be the primary aim. Rather the ability to live with other people in a certain minimum degree of independence is the desideratum.

Therefore, although individual sessions should and must take place, a blind child will have to be guided step by step into the group and must become part of it, whether it be the family or the particular grouping at school. He must be included in meals, routine, pleasures and work in his own way, so that he experiences that he is part of a whole and is not condemned to utter isolation.

And finally, if therapy is to be based on human relationship, by the reciprocal nature of things it is of equal importance for the therapist, indeed for all concerned, to regard a blind and deaf child as part of their own existence in spite of the isolation his severe handicap imposes.

If those who live and work with such children are prepared to train themselves to become sensitive beyond the level that their ordinary conscious mind requires, along the lines briefly indicated in the chapter on Child Development, they will learn to live with these children in a meaningful manner. Personal development and progress will provide the necessary environment for the development of children as severely handicapped as blind/deaf children are.

I shall be saying more about the importance of the relationship between the environment and the development of the handicapped child in the final chapter of this book.

7. The Child with Impaired Hearing

It is likely that early efforts to teach deaf-and-dumb children to read, write and speak, which started in the sixteenth and seventeenth centuries can be regarded as the historic beginning of special and remedial education. Thus, at the beginning of the last century, the deaf had that significance which the cerebral palsied now have in our own, when it was discovered that in spite of their severe handicap, intelligence can be intact and even fairly high.

Deafness can occur as an inborn handicap in children in the form of an isolated defect, and occasionally such children will learn to lip-read and to speak to an extent that their deafness passes nearly unnoticed, but unfortunately this happens rarely, even though over the past twenty years or so there has been an increasing awareness that deafness in children is hardly ever total and that, therefore, early stimulation of residual hearing can lead to a considerable use of available hearing potential.

Here, however, the concern will be only with those conditions of deafness which present a fundamental developmental problem to the child. In order to feel one's way into the developmental aspect of deafness, it may be helpful to consider deafness developing in a normal adult person.

As deafness increases in an adult, he not only becomes more isolated but often also suspicious of others. This can be readily understood if it is seen that the deaf person continues to *listen* and expects to hear things, but as his hearing is increasingly poor, he either does not understand what he hears or altogether cannot hear what is said, and because of the intensity with which he listens and expects, he may imagine he has heard things which in fact have not been said.

It becomes obvious that in the total complex of hearing, there are two fundamental elements—*hearing* and *listening*. Hearing is clearly a primary sensory ability, while listening is a human *activity*.

When a normal person of advancing years gradually becomes deaf, his hearing ability is affected, but not his listening ability. This can, however, be very different with the developmental aspect of deafness in children, whether the deafness is congenital or sustained later. Therefore the nature of the deafness must be scrutinised, and components pertaining to *hearing* and *listening* must be clearly separated and distinguished.

To begin with, it will, of course, be held that the appropriate measure to take to help a child's *hearing* problem is the amplification provided by a hearing-aid. For the adult person, this is in most cases a suitable measure, but it is not always as successful as might be hoped when it is applied to handicapped children.

The use of the hearing-aid presupposes that the listening ability is intact, and in fact, it requires a considerable degree of discrimination to make use of a hearing-aid, and some children suffering from congenital deafness do not appear to have it to a sufficient degree.

There is, in addition, just reason to hope that sufficient hearing stimulation may by itself stimulate the ability to listen. However, again with children, this does not always happen and the problem remains as to what other means can be employed to encourage listening activity in a child.

It is tempting to do this by visual means and, no doubt, a child's interest and attention can be won by doing so. Yet, the unfortunate "marriage" between *looking* and *listening* is to be noted here, and it will be seen that there is a degree of competition between the two. In looking, there is an inclination to listen less intently, and in listening, the ability is enhanced if *no* looking takes place and the eyes are shut.

Some first-hand experience may be gained of the actual activitity or quality of listening, when, for instance, one sits in a quiet room where a clock is ticking and makes oneself aware of the ticking, and then lets the ticking recede from one's hearing. One finds that one can learn to hear the ticking or not to hear it at will, and thus conscious control is gained over one's own listening power. It may then become clear that the activity of listening is concomitant to an outspoken muscular relaxation and stillness, in which all activity that would otherwise go into limbs and muscles or into the act of looking, is then directed to hearing.

Perhaps it may now become clear that the developmental problem in a child with a defect in the realm of listening is a twofold one. He may either not possess the necessary activity and initiative for concentrated attention and listening, or he may be too hyperactive in other ways, and thereby incapable of achieving the degree of muscular relaxation that is essential for initiative and activity to be directed inward to hearing and listening.

Two types of deaf children are encountered. Those who are always "on the go", over-inquisitive, restless, jumpy, nervous, and those who tend to be lethargic, slow, placid, inert. These two types can either appear in relative distinctness, or in combination to varying degrees.

In some cases, the hyper-activity and restlessness in deaf children is due to a slight athetoid palsy. Such children may have suffered from jaundice of the newborn and thus may have sustained some damage to the nervous system, which causes muscular nervousness and twitching and aggravates the customary restlessness of deafness. In such cases, the use of the hearing-aid is not only unsuccessful, but it even increases their restlessness and athetoid tendency, and thereby also their deafness.

Similarly, it can happen that a child who, to all appearances, has an impairment of hearing, is actually suffering from aphasia, which means that his *hearing* of sounds is not impaired but rather his perception and recognition of speech sounds and words or, in other words, of language is impeded. This condition can in some ways be compared to cortical blindness and will be described separately and fully in the next chapter.

In such cases as these, the amplification provided by a hearing-aid will give increased sensations of hearing but it can in no way improve the perception of words or language. On the contrary, the child may misinterpret and think that he is expected to learn to enjoy the particular noises that may be produced by acoustic feedback in the hearing-aid and will be tempted to use it to provide for himself a kind of electronic music and, therefore, his problem is aggravated.

I do not, however, wish to be misunderstood as standing in general condemnation of the use of hearing-aids for children. On the contrary, if a child who has an isolated ordinary defect of hearing is not provided with an adequate hearing aid, he can be prevented from learning to hear and to speak in a way that will prove disastrous for him. I only wish to point out that the hearing-aid may be used only in appropriate circumstances with care and in full knowledge of what is required.

The serious impairment of listening and hearing has fundamental and far-reaching consequences for a child. The first and most obvious is the inability to gain experience and mastery in the realm of language and thus being cut off from communication. But not only this—the development of the experience of *concept* in general can be profoundly impaired in children suffering from congenital deafness.

The actual perception of words and language as well as the perception of concepts on a sensory basis, shall be described in full in the chapter on Aphasia to follow. Here, however, only the fact that the experience of concepts and especially of abstract concepts becomes very difficult when the means of learning for a child is mainly visual need be highlighted.

I shall give an example. A child is shown a chair. He is then shown the word c-h-a-i-r and learns to write it down, having learned that this word and that chair represent the same thing. But one day he sees a green chair and does not know what it is. For all his previous labour, he has not been able to build up the concept "chair". This is a simple example but one that has actually occurred, and serves to point out what the precise problem is.

The child who is cut off from language and who, as a consequence has the greatest difficulty to form ideas and abstract concepts is in danger of an indiscriminate emotional and moral development.

Listening and hearing are gateways into human cultural existence, and it is therefore of the utmost importance that every possible effort is made to help a congenitally deaf child to achieve some experience, however rudimentary, of the worlds they open up.

The importance of continued speech in the presence of deaf children has already been spoken of, in spite of the fact that their muteness inclines to frustrate speech in others. The hearing of one person is partly contained in the speaking of another, and therefore speech has a certain power of calling up hearing and listening in the other. This constitutes a tool that should be used to the utmost. Hearing stimulation via speech should, in order that the limited potential of a deaf child is not overtaxed, make use of a limited vocabulary which avoids synonyms or homonyms, so that latent speech development is not unnecessarily frustrated.

However, in the case of a severely deaf child, the use of speech in itself will not produce sufficient stimulation for the development of hearing and listening. In such cases, direct hearing therapy is called for. For instance, a strong, loud tone may be sung into the ear of the child, who is then helped to strike the same pitch by using his own voice accompanied by raising or lowering of the arms according to the pitch of the note. By these means, a child can be taught to appreciate the pitch of tones and intervals between them both in his hearing experience and in his own vocal production. Even though he may be too deaf to hear speech at all, these first steps are of the greatest value, for they constitute a breach of the wall of deafness that surrounds him.

Such primary hearing exercises as these can be used to lay a foundation for the experience of communication, and can best be done with a small group of children who sit in a circle with the therapist, who sings a tone into the ear of the child nearest to him and helps him to produce the pitch in the way set out above. This child then passes the pitch on to the next child, whose tone production may have to be corrected, both by the therapist and by the first child, but in this wise, the tone passes round the whole circle until it returns to the therapist. Practised in a group, the hearing stimulation and perception of the children is not only linked to personal tone production but equally to a basic form of communication, which will be valuable in the subsequent development of the deaf child.

Tone production in deaf children must be supported by a variety of means, such as feeling the therapist's throat and their own whilst speaking or singing These same principles, with the group work described above, are again applied when it comes to actual speech therapy. Therapies must, of course, be carried out by experts and the precise details of their work is outside the scope of this book.

It is now generally recognised that deaf children should not be exposed to unnecessary and avoidable noise but that, on the other hand, it is of considerable value to them to attend concerts if they can sit where the volume of sound is good. If given the right opportunity, they can, to varying extents, learn to appreciate music.

In intensive speech therapy, which must, of course, include visual and phonetic teaching as well as every possible auditory means, it is of great importance to give a child sufficient emotional and moral stimulation in addition to the stimulation of abstract concept. He may be quick to grasp the superficial nature of things and their overt manifestations, but the profounder meaning and significance of things initially occupies alien territory into which he must be guided with great care and understanding. His religious or moral education is for this reason a fairly problematic matter. Dramatic performances and puppet-shows, if primarily designed to impart meaning rather than semblance, can be of the greatest help here. These will also help him to learn to love what should be loved and to abhor what should be abhorred when it comes to "right and wrong" in dramatic, human situations, applied ultimately to himself and his own actions, for as was mentioned earlier on, the congenitally deaf child tends to be morally and emotionally indiscrimate and is thus endangered.

It follows that the entire realm of the arts is of vital importance to the therapy and education of deaf children, not only because by means of the arts things can be conveyed to a child which cannot be conveyed by language, but also because he himself needs a means of expression, and here the arts are eminently suitable as a vehicle or medium. Some of the most severely disturbed deaf children, particularly those in whom deafness is combined with autism or psychosis, show outstanding gifts in the visual arts, just as blind children do in music.

Through artistic expression, they learn to know the meaning of things and can make themselves known to and understood by others.

8. Children Suffering from Aphasia

While "blindness" and "deafness" are self-explanatory terms in the sense that everyone knows what is meant by them, it will require a brief introduction to be able to use the term "aphasia" with equal clarity of meaning.

Basically, two forms of aphasia can be distinguished; the so-called sensory or receptive aphasia or auditory imperception, which is an inability to perceive words and language while hearing itself is unimpaired—or, executive or motor-aphasia when a person can hear, perceive and understand speech and is yet unable to speak, not because of a faulty or damaged motor-speech organisation in the sense of palsied speech muscles or palsy of the larynx or vocal chords, but because of a more basic inability to speak.

Medically, such conditions are not caused by impairments of the ear or of the speech-organisation, but by disturbances and impairments in certain areas of the central nervous system. A somewhat similar or allied condition is described as Agnosia where there is an inability to grasp the meaning of words or things, although the intellect is otherwise normal. This condition too is connected causally to certain forms of damage to the central nervous system.

If, however, the significance of these disturbances and conditions is to be experienced, the more basic interpretation and understanding of sensory perception generally must be recollected, as the conditions of aphasia are fundamental disturbances or impairments of perception.

Obviously the eyes are used for seeing, and the ears for hearing. It is not so obvious that in the act of seeing, two separate qualities are involved which belong to two different parts of the sensory organisation. One quality, the perception of colour, is directly related to the retina. The other, the perception of form and shape, is part of the sense of movement related to the muscles of the eye as well as to a set of pathways to the central nervous system different from that used in the perception of colour.

It may also not be a matter of full awareness that actually only one definite point may be looked at or focused upon at one time. For example, when looking into a mirror, it is not possible to look into both reflected eyes at the same time, but only at one of them. When it is required to focus sharply, it must be done on a single point.

Hence, it may be realised that perception of the form and shape of an object is only possible to the extent that the eyeball moves about and follows its outline and contour, something which is not at all the case in the perception of colour. Thus, these two sensory elements are distinguished at work in the sense of sight.

The sense of hearing may not be scrutinised in a similar way. When a group of people is trying to learn a song, most of them will have retained something of the melody after having heard it once. They will not, however, have retained the words. They may, in fact, have not heard the words at all to begin with, only the music, and the tones and intervals in certain rhythms. This is a fairly common experience.

But there is another corresponding experience of which the awareness is usually not so acute. A poem say, is recited. Afterwards, most will have retained the gist of the words, but will not be able to recollect the tonal melody of the voice that spoke the poem. Days later, the words will be remembered but the quality of the voice of the speaker will defy reconstruction in the memory.

On the basis of these two simple observations, it can be stipulated that the sense of hearing contains two elements: the perception of melody or tonal qualities, and the perception of language or words. The acoustic principle involved in the difference between these elements of perception is the difference that exists between tones and overtones. All tones with the exception of some electronically produced ones have, besides their pitch-quality, a number of so-called overtones—upper partials or secondary tones that swing and sound with the pitch tone. These overtones play a decisive part in the make-up of the quality of speech sounds.

The distinction of pitch-over-tones can be noticed when the same tone is played on different instruments. An ordinary person distinguishes whether a tone has been played on a violin, flute or a piano. There are people, however, who find it difficult to distinguish the instruments on which a tone is played.

It is interesting and indicative that most of those children who suffer from a degree of auditory imperception or sensory aphasia, although their hearing is acute, cannot consistently assign the same tone to the instruments on which it is played. The ability to perceive the spoken word and language as different from all other sounds is based largely on the ability to perceive overtone qualities exclusively while suppressing pitchtone qualities from consciousness; in the hearing of music however the perception of pitch is dominant.

For the appreciation of aphasic conditions in child development, it is essential to observe how the faculty to perceive overtones develops in a young infant. It is obvious that in a normal child, hearing begins to function at a very early age. Already a few days after birth, an infant shows distinct reactions to sound and noise. However, it is equally obvious that his reaction to the spoken word is no different from that to other tones and sounds. There may be a reaction to different tonal qualities, but the phonetic quality of the spoken word cannot yet be experienced. It is only towards the end of the first year that a new perceptive ability begins to develop and the child first perceives words and then forms his own first words. The awakening of a new sense at the end of the first year of a child's life can to all intents and purposes be spoken of as a sense of word and language. Thus a new perceptive sense-quality is added to a child's sensory experience.

A further self-observation which is more striking and evident than our observation of earlier on is helpful to an understanding of agnosia. Let us say, an interesting lecture delivered in an international multilingual conference has been heard. Upon returning home, the content of the lecture may have been fully and accurately retained, but the language in which it was delivered is beyond recall. Equally, it is a rather general and obvious fact that, after listening to an interesting conversation, the gist of it can be fairly accurately recalled without the words being retained.

Once such diversities in the experiences of hearing are noted, it will be sensed that, in our total perception, there is a domain for the perception of meaning, content and thought separate from the perception of language. In other words, over and above the sense of hearing which gives us the perception of tonal qualities, there is a separate perceptive ability related specifically to words and language, and over and above this in turn, there is a further perceptual faculty which allows the perception of thoughts and meaning. In fact, the sense of hearing is revealed as a "gateway" to two further, and, one might say, higher senses—a sense of word, and a sense of thought or concept.

As the sense of word seems to unfold in a normal child at about the time when he achieves uprightness and walking at the end of his first year, the sense of thought or concept begins to be effective around the end of the second year after a child has acquired language.

It would seem that the acquisition of the sense of word is the outcome of the achievement of motor-control expressed in uprightness and walking, while the sense of thought or concept is the result of the second phase of motor-descent which is the incipient use of language.

In Chapter V on Autism these two great developmental steps were seen as the precursors to the dramatic event that comes between the second and the third years, when a child first experiences his own ego. Here is seen the connection between perceptual disorders and childhood autism and it can be understood why childhood autism is sometimes mistaken for perceptual disorders of an aphasic or agnostic nature. It must also be considered, of course, that a child suffering from a severe degree of sensory aphasia or agnosia, or even both, may be forced into a situation so intolerable that a panic reaction similar to that found in autism may result. In fact, there are many children suffering from aphasia and agnosia who present typical symptoms of childood autism or psychosis, but most of these children can be helped to overcome their autistic or psychotic features if their basic aphasic condition is recognised and to some extent alleviated.

It becomes equally clear why deafness and degrees of aphasia can be mistaken for one another, as both conditions make auditory perception difficult or even impossible and both are obstacles to the development of the sense of thought.

Unfortunately, conditions of developmental aphasia and agnosia are not always easily discernible. Children who do not develop speech, who do not seem to understand the spoken word and who cannot grasp concepts, are understandably regarded as being severely retarded or mentally handicapped. They may appear to be restless, hyperkinetic, and are frequently aggressive or negativistic, because of the frustrating situation imposed upon them by their underlying aphasia or agnosia.

As developmental aphasia is caused by some form of brain pathology, it may be accompanied by other results of brain damage, which can lead to an alternative diagnosis, allowing the element of aphasia to pass unnoticed. As has been said, the symptoms may be similar to or identical with those of autism or psychosis and a child is then diagnosed as autistic or psychotic.

Once, however, the nature of the aphasic or agnostic problem has been understood and one becomes sensitive to the potential of handicapped children, it is possible to distinguish between avoidance of communication in the autistic child and the inability to communicate in the aphasic child, and to distinguish between the inability to grasp concepts in the agnostic child in spite of his innate intelligence, and the impaired ability to think in the retarded child.

The basic requirements when confronted with any type of handicap or disturbance in children is, as has been seen, the adjustment of personal attitude. It is this adjustment that establishes the premise for any further therapeutic effort. The restless, hyperactive, destructive behaviour of the aphasic child can arouse an attitude of defence on the part of those around him, but just this attitude of defence blinds our eyes to the motor-intelligence displayed in the very acts of destruction performed by the child.

We have observed a little boy who displayed unusual inquisitiveness by opening ladies' handbags, pockets, and drawers and taking out all the contents. He seemed to choose to do so especially in the presence of others and even more markedly, in the presence of the actual owners. He would go up to a newly arrived guest and open her handbag while it hung on her arm and have the contents out in no time.

Understandably, such behaviour is interpreted, to begin with, as a lack of manners or restraint or also as maliciousness and this is what naturally calls up the attitude of defence. But on closer and more sympathetic examination, the behaviour of a child suffering from aphasia and possibly from agnosia as well is the expression of his continuous frustration of communication and a demonstration of his eagerness to display both his intelligence as well as his active interest in his surroundings.

There are other children suffering from conditions of developmental aphasia who present themselves as partially withdrawn but who will suddenly dart out at another person emitting an intense and sharp bellowing noise. For months and even years, it may go unnoticed that this child is trying to make use of the one word he has managed to perceive and reproduce. Again, our own inability to interpret such things correctly is more often than not based on the defensive attitude that such abrupt and unusual behaviour illicits.

Once everyone in the environment of a child with aphasia or agnosia has become aware of his conditions and realises his other relatively less impaired potentials, both hyperkinetic symptoms and autistic features will recede, which eases the child's own situation considerably because his genuine problem has become manifest. Direct therapy can now begin.

Therapy will consist of various means of assisting the child to distinguish first the overtone qualities in the same pitch-tone rendered by the human voice as well as by a selection of musical intruments, as we have indicated above.

The next step is to introduce words by careful phonetic sounding and by lip-reading. Phonetically spelt words and pictures of the objects they depict together with the name of the object as it is ordinarily written are used right from the start. Eurythmy gestures for each of the vowels and consonants are an invaluable support to the child's development of sound- and word-perception. The effective use of all these elements requires special training on the part of the therapists, and must be maintained over long periods.

Children who suffer from complete auditory imperception, or in other words, total receptive or sensory aphasia, can in the course of a number of years learn to perceive and understand simple speech as well as to express themselves, if only in a laboured and halting fashion. Others may only be able to acquire the very first rudiments, while a few may occasionally nearly normalise their word-under-standing and use of language.

It is not unusual for children suffering from severe degrees of aphasia to go through prolonged periods of frustration and aggression in connection with their therapy sessions. Having previously become somewhat adjusted to their incapacities, treatment now upsets this adjustment and new demands have to be faced for which the child is naturally not equipped. Situations such as these require great tact and empathy on the part of teachers and therapists so that frustration remains meaningful. In any case, the steps to be taken must be of such a size that a child achieves success within a reasonable period. Needless to say, any form of dramatic performance in which speech is supported by movement, gesture and mime, either directly on stage or with puppets is of the greatest therapeutic benefit to children with aphasic conditions.

In cases where speech is more severely affected than can be accounted for by the degree of auditory imperception, and, of course, in all cases of predominantly executive aphasia, a different line will have to be taken in finding exercises which will prove remedial.

In the chapter on Dominance and Laterality, that is, on the Left–

Right problem, we outlined how motoric speech development is linked in a certain way to the establishment of dominance. Children whose motoric speech development is impaired in the sense of executive aphasia frequently show signs of ambiguity in dominance. In such cases, it is not only valuable to correct any possible crossed-dominance, but to develop dominant unilateral movement in derivation from rhythmical bilateral activity. For example, in running, both legs are used in rhythmic alternation and the activity is bilateral. Then may come a long or high jump in which one leg has to lead and the other to follow and in which dominance is the important element. The development of unilateral dominant movement out of bilateral movement can be effected by running and jumping or ball-throwing and other exercises, such as javelin- and discus-throwing. Although the latter like the former, are fundamental to the development of speech, they should not be attempted with a more handicapped or retarded child because of the demand on relatively unimpaired motor-co-ordination.

There are cases of true motor or executive aphasia where a child has relatively good word-understanding but fails to acquire speech, even though he has no paralysis of the speech muscles. These children often manifest their lack of dominance-development in a striking and impressive way. In their general movement pattern they show pronounced bilaterality in that they walk with a particularly broad gait, each leg tending outward towards its own side, with the arms often spreading sidewards too, which gives the impression of their walking with the left half of the body to the left and at the same time, with the right half to the right. A spread-eagle posture appears in this peculiar kind of gait which is enhanced by a tendency to open the mouth wide in an attempt to communicate by speech. Many of these children can produce sound but are incapable of forming articulate words. Just for these in particular, the exercises described above are of special importance and can sometimes bring about an initial degee of speech.

While it is a gratifying experience to help an otherwise relatively normally-developed child with an impairment of word- or language-understanding or impaired speech, it is a very different thing to be confronted with a child whose entire development is marred and hindered by a problem of aphasia, and then to see the developmental changes that can be brought about, once his aphasic condition is understood and therapy started.

As a result of the attempts to interpret child development detailed in this book, it can be appreciated that the development of the sense of word—the development of heard, understood and spoken language —is one of the most fundamental stepping-stones in early child development, particularly with regard to a child's ego-integration. In children who do not develop the sense of word at all, who suffer from so-called total aphasia, there is often a severe and widely ranging failure in development.

As was mentioned earlier on, these children are frequently not diagnosed as aphasic, but rather as severely mentally retarded, with little chance of further development; the autistic and negativistic features which accompany the condition of severe aphasia and which may lead to wrong diagnosis have been described.

From conditions of nearly *in*human distortion of behaviour such children can be regained for human society by being given the chance to experience a first dawn of communication with others.

Just as for the blind or deaf child it is of existential value to achieve a first experience of light and sound respectively, in order to open up for them the worlds of visual space and time, it is of equal existential value to the severely or totally aphasic child to achieve the basic experience of the word and its communicative and contact-building significance.

Through this, the child can be led to an encounter with the human spirit which is the ego in his own existence and the ego in the other person.

To lead a child who had been excluded from this experience into a participation in it is one of the most fulfilling things a man can do.

9. The Emotionally Disturbed and Maladjusted Child

A great deal of importance and value has been written about emotional disturbance and maladjustment in children and adolescents. I shall confine myself here to two aspects which, though relatively little known and described, have played a significant part in my experience with disturbed and maladjusted children. One is an aspect of physiological-psychological constitution, the other of environment.

It would probably be considered that children who fall into the group headed "emotionally disturbed" or "maladjusted" are those whose general and intellectual development are relatively unimpaired, but who nonetheless show modes of behaviour and reactions to situations that are unusual and often offensive and hurtful to others, or even to themselves.

To what extent this different or unusual type of behaviour is the result of a different way of experiencing things is for the most part not quite clear. A positive trend in customary descriptions is found in the assumption that disturbances in emotional reaction and behaviour are environmentally caused and are the result of hurtful or traumatic situations in earlier upbringing. This is a socially and therapeutically valuable attitude since it is usually easier to do something about the environment, than to alter the constitution of the child in question.

Nevertheless, we must be aware of the fact that any type of developmental impairment or handicap such as those described in the previous chapters can play a basic part in the child's peculiar vulnerability and mode-of-reaction.

There is, however, yet another factor, so far not mentioned, which may play a more specific part in the constitution of emotionally disturbed and maladjusted children. In an earlier chapter, we described certain form-qualities or form-polarities, such as large-headedness and small headedness, integration into left and right, into space and time, as fundamental to child development. There is another rather different constitutional polarity to be considered, especially when emotional problems arise in childhood. It can best be described in the following way: A person may feel his own body, his organs as something that restricts or imprisons him, that forms a barrier between the rest of the world and himself. He is inclined to be insensitive to his surroundings, but is easily offended, inclined to be aggressive because of an irritability with himself rather than with others. On the other hand, some people feel exposed and unprotected by their body; their surroundings perpetually threaten to impinge upon them. They feel as if their skin were raw, as if they did not have enough skin. Not only the physical vicinity of another person but even his words get "under their skin". He is vulnerable, over-sensitive and constantly irritable.

These two types can be rather well defined particularly amongst children. An over-sensitive child will react badly to any change in environment. When he is taken to a new school or locality, he will be inclined to withdraw, to fuss, to be unhappy and it takes him a long time to readjust and settle in the new situation. But if he stays in the same surroundings for a sufficiently long period and does not have to meet new people he will gradually establish himself and make good progress.

It is quite a different matter when it comes to the other type of child. To all appearances he will be greatly relieved when he goes to a new place and will seem to be relatively happy, adjusted, co-operative and helpful. But as time goes on, and he becomes familiar with the new setting, things begin to lose their newness and no longer hold his interest. He begins to be faced with the fact that he himself has remained the same in spite of the new surrounding and this is the one thing he cannot bear, and with which he cannot really live.

This type tends gradually to become difficult the longer he is established in a new place, and he elicits the aphorism—a new broom sweeps clean which is a pathetic misinterpretation of his situation. This kind of reaction may well be based on his constitution which gives him the sense of being imprisoned in his own body—tied, as it were, to himself—a negative, self-destructive state of which he can only be relieved when there is sufficient challenge from outside, in something new that enables him to get beyond himself.

In contrast, the over-sensitive type of child will be inclined to feel that any demand made by others far exceeds his own potential and is therefore over-powering. Thus, his most frequent exclamation is: "I can't." Nevertheless, these children often possess excellent abilities, artistic and otherwise and can distinguish themselves in surprising ways, simply due to their over-sensitive and hysteric constitutions. They can direct situations as they wish because their extraordinary sensitivity allows them to foresee and guess the intentions of others. Therefore, they are past-masters in playing one adult against another, not because they are malicious, but because they are too open to and over-involved in the subtle interplay between others.

Typical hysteric behaviour patterns only become understandable when seen as the result of the unusual ability to generate situations on the basis of exceptional sensitivity. This sensitivity is usually organically determined and of the type described at the beginning of this chapter.

Pictorially speaking, we might say that the body is relatively frail in relation to the force and magnitude of the wishes and emotions it must contain. Consequently the organism yields readily to the impulses of what must be called the *soul*, though by no means voluntarily.

Children who possess this kind of constitution to an extreme can from one hour to the next display acute symptoms of various illnesses, including epilepti-form convulsions and the like. It would be a misinterpretation to think that they are shamming. Rather, these manifestations are caused by the power exerted by the emotions over the organic functions. Such extreme cases are somewhat rare and I have only mentioned them to characterise a constitutional trend in some emotionally disturbed children.

Children who become disturbed on the basis of an hysteric, over-sensitive constitution, can be helped in a number of ways. An important facet of their education and therapy is to help them to achieve a more elastic relationship between their desires and emotions and their bodily organism. A very simple and yet very effective thing to do is to provide for such children all manner of exercises in which activities as diverse as writing or walking or going up and down the stairs, are begun slowly, then gradually increased in speed to reach a peak, after which the speed is gradually decreased to the slow pace with which they began. By doing these exercises repetitively, a child subconsciously readjusts the relationship between his mind, or "soul", and his body.

Children with outspoken hysteric conditions require a considerable degree of mature and sound human warmth from others. It does not help when those concerned believe that extreme and dramatic symptoms of this or that kind are *simulated* by these children. The adult must accept the symptoms as real, because the child actually suffers them, notwithstanding that they may not be organically triggered off, but there is instead a considerable emotional involvement. This must be understood before any effective help can be extended.

It is astonishing how difficult it is, even for the medical profession, to abstain from regarding hysteria as simulation. Yet the ability of some eastern Yogi to walk bare-footed over beds of glowing embers, is based on a similar interaction of psychological and physiological factors to that in hysteria. Here, neither the coals nor the walking on them is simulated.

The other type of condition which is described above—that in which a child feels caught and imprisoned in his physical organism, is more difficult to alleviate. In extreme cases, it manifests in epileptiform convulsions and these can be seen as a drastic attempt to fight free of engulfment by the body which is the result of some organic pathology.

After a period of sleep, these convulsions lead to a noticeable degree of relief and freedom, but this is usually temporary. The aggressiveness of these children which is not primarily directed against others but against their own inability to be free of their organism, presents a very great problem.

When we realise that this aggression is organic in origin, it may have to be discussed openly with particular children. In spite of the fact that their situation is constitutionally determined, they must learn to acquire a responsible attitude towards it. A basic principle in the moral education of such children is never to condemn *them* but only their *actions*, and then their actions only when they are fully reassured of the continued warmth and sympathy of those who are concerned with them.

Again, fundamental therapeutic exercises exist to ease the burden this type of constitution imposes on a child. Most children of this type show some impairment in their adjustment to the gravitational field, and a widely varied selection of balancing exercises, trampoline jumping, and exercises in spatial orientation through gym and other means provide specific support.

For children in whom the fluid element becomes unbalanced, who vomit during their fits or who are more prone to convulsion when they have eaten and drunk much, swimming and exercises in water are most valuable and beneficial. When the difficulty seems to lie rather in the air or warmth systems of the body, appropriate exercises which make specific use of these elements are indicated.

It is interesting to note that these two polar conditions of over-sensitivity on the one hand and engulfment by the body on the other can be discerned through a simple observation of the hands. In the over-sensitive, hysterically inclined child the palms and fingertips will form the predominant part of the hand and will seem to outdo, as it were, the back of the hand. The hands are usually moist and there is a general proclivity to perspiration. The fingertips tend to overgrow the nails, which are usually soft, short and frequently bitten.

The other type of child has a hand in which the dorsum, the back of the hand, is the predominant part. The hand is on the whole firmer and harder, and the nails tend to curve over the tips of the fingers. There is little tendency to moisture and the whole hand feels less open.

The hand is, of course, only one manifestation of a much more general constitutional type, but no less indicative for that. When one has a certain sensitivity for constitutional types—and this can be trained—the two types of children just described can be readily distinguished. There is the soft, open, somewhat blooming, promising, youthful-looking, hysteric, over-sensitive child, as opposed to the hardened, denser, disappointed-looking and self-contained appearance of the engulfed, epileptoid child.

It is obvious that extremes in either of these two types of constitution will tend to make a child vulnerable, and prone to breakdown in unfavourable environmental circumstances. But at the same time, of course, any of the other handicaps that have been described will be contributory to emotional disturbance in children who are exposed to environmental stress or damage. The effort must always be made to discern whether maladjustment is purely environmental or developed on the basis of a constitutional frailty or a more basic impairment.

In any case, the emotional distress and maladjustment of a child will always have to be dealt with on an environmental basis, whatever else is done for him. He will require years of sustained sympathy and warmth without the expectation of overmuch response from him.

Emotionally disturbed children seem to be under the spell of self-abnegation and the desire to hurt those who love them. From the point of view of their emotional development, this can be readily understood as a reaction to hurtful experiences in early infancy.

It is rarely possible to achieve during adolescence complete adjustment in a severely hurt child, but according to my experience, if sympathy and love for him can be sustained for long enough, even through the difficult period of adolescence and early adulthood, adjustment and relative harmony establish themselves later on.

It would seem that the core of the problem of the emotionally disturbed and maladjusted child is that he himself does not succeed in learning to love and thus to take responsibility.

The ability to love is probably rooted in early infantile experiences. To some extent this has been described in Chapter Two on Child Development, where, towards the end of the chapter, motivational development and the origins of personal morality were linked to infantile stages of experience. Wrongs and deprivations in this early phase of development can hardly be retracted and the wound can rarely be healed. In *The School as Environment* in Chapter Four, there is a description of attempts to provide during a later phase of child development that which should have been present earlier.

Here I want rather to describe a specific moment in child development which is particularly relevant to maturation in the sense of accepting responsibility. It is a moment which in an ideal childhood immediately precedes or coincides with the onset of adolescence. A boy begins to sense that his father is after all not really so awe-inspiring, is in fact scarcely worthy of being the father of such a son, and yet at the same time he knows that his own conduct, his attitude towards him, will make him a worthy and valid father. Obviously the latter part of this experience is the salient one. Where does it originate? It arises, at least to some extent, from a child's realisation that his behaviour, his treatment of others is of actual support to their existence, an experience that can only arise when a child meets another in a need he can help under ideal or, at least, favourable circumstances. Such encounters can happen between an older and a younger child, between a child and a frail old person, a boy and a girl, but most readily between a healthy and a handicapped, or a bright and a dull child. The maladjusted child has only too often an intense longing to help, to succeed, to have something to give, to be the hero. If he lands in a segregated school exclusively for maladjusted children, opportunities for such experiences are wellnigh eliminated. The only realisation of his positive longings are not to tell on others, not to fight or to steal, not to outdo the others in pranks. It is an interesting observation that severely disturbed and maladjusted youngsters have occasionally improved considerably when finally admitted to adult psychiatric wards having previously been unmanageable in all other settings. This is due to the fact that among frail old psychiatric patients they have valuable help to give, especially in the usual situation of shortage of nursing staff. For the first time, they are in a situation where they are genuinely needed and thereby the first step towards self-realisation and responsibility may be taken. Not that, of course, this is intended to advocate such unfortunate measures as admission to psychiatric wards, but merely to serve as an illustration of the particular beneficial, therapeutic situation for maladjusted children. If emotionally disturbed and maladjusted children are to learn to love and to take responsibility, they cannot do better than to grow up in an environment in which they encounter children who suffer from other problems and handicaps. Particularly, the presence of a physically handicapped or severely retarded child can be healing for the maladjusted.

The interaction between children suffering from different handicaps will be gone into in the last chapter of this book.

10. The Education of the Retarded Child

Many handicapped children are noticed for the first time because they are unable to progress in school as quickly as their peers and when they begin to fall behind. The so-called educationally subnormal child is not referred to here, that is the normal, healthy child who fails in school because the programme is geared to above-average or average intelligence, whilst his own intelligence is below average.

The subject of this section is the truly handicapped child whose development has been hindered or marred in one way or another. In the school situation, a teacher may, to begin with, have the impression that a particular child is not able to follow and to advance abreast of the other pupils in some subjects. He will, therefore, be inclined to suggest that the child have extra coaching lessons in order to catch up.

This, however, although well meant, may prove not only fruitless, it may aggravate the situation. At the same time, the parents will have noticed that their child is not only wanting in this or that achievement, but that he is making little progress, altogether that he is not developing as might have been expected at school.

The child himself may have become somewhat reticent, irritable and possibly aggressive. His behaviour will be difficult and negativistic, or he will be inclined to become timid and depressed.

Behind the different views of teacher and parent, there is the general problem of a twofold aspect of education. There is, on the one hand, the element of actual *learning*, the absorption of information and the acquisition of abilities and skills. But on the other hand, there is an aspect upon which scant emphasis is laid in modern education the development and maturition of personality.

It is by no means only the first of these aspects which is impaired in a handicapped child. The problem lies more often in his growing up and maturing as a person. Indeed, just here lies the distinction between the normal, healthy, but constitutionally dull child and the handicapped child whose problem is a developmental one. Here again, the question which was discussed in the introduction to this book that of remedial education as over and against special education appropriate to the child's measurable intelligence is encountered.

Both aspects of learning and developing are, naturally, closely interrelated, but here they will initially have to be approached separately. The processes of teaching have to be designed with particular regard to a child's handicap. They must be adjusted to his specific abilities and disabilities and to the particular nature of his developmental problems. I shall return to this question later on.

The developmental aspect of a handicapped child's education must be seen on a wider basis not only because a child's need in this respect is basic in child development as such, but because it becomes more outspoken, more urgent, more precarious a problem for a handicapped child, whereas a normal child maintains his course of development even in relatively unsuitable circumstances.

The first essential support to be given to a handicapped child consists in enrolling him in a class of his contemporaries, some of whom will progress faster, and others slower than he. If, because of his retardation, he is included in a class of children younger than himself, as easily happens on the basis of so-called mental age, he will be held back in his general development and suffer additional hindrance to the process of maturition. (Here, however, an interesting observation has been made—handicapped children who are physically very underdeveloped and under-sized often respond well to being placed in classes of children much younger than themselves, where they begin to catch up both in physical growth and in general development.)

In a class comprised of children of the same chronological age, but with widely differing abilities and disabilities, subject-matter cannot be selected according to what can be learned best and quickest by a particular group. It must be selected according to the developmental needs of the children, so that the teaching programme accompanies or conducts a child on his way from the childhood world of fantasy to the reality of adult life.

Taking up a child's earliest experiences of himself as the Maker, the Creator, in the various activities in nursery class age, going via the fairy-tale through myths and legends, a child is then introduced to the archetypal events of the Old Testament, which marks a departure from the pure world of fantasy and an entry into a world of morality and the beginnings of human history. Then, by the age of twelve, a child is ready for the transition from the Greek to the Roman phase of history, and so on continuing in this developing fashion into the upper classes.

Some severely retarded children in a mixed class of chronological age will not be able to grasp what they are being told. Others will grasp its significance fully, and if a subject is taken for a period of weeks, the children will begin to play "Greeks and Romans", for example, in their playtime, when more limited children are drawn into these play-activities and experience these fundamental phases of human development, to the great benefit of their own individual developments. The histories of the Middle Ages and the dawn of contemporary times are suitable supportive subject-matter for children approaching puberty.

The choice of particular sciences and other ancillary subjects is also determined by the original developmental process in education. Therefore, it is appropriate to introduce scientific subjects as such with acoustics, developing the concept of proportion in relation to musical intervals as it happened in early Greek times. This may then lead over to optics as an interim subject and finally, when the child approaches puberty, to mechanics.

In this way, a child's gradual "descent" from the imaginative to the mechanistic appraisal of the world is done proper justice, and the children learn, not only in an intellectual fashion, but also creatively and with intuition, which supports the maturition of the personality, and again augments the ability to learn.

During the first seven years of schooling, the aesthetic argument, the argument of the beautiful is mandatory. Only at the onset of puberty should the argument of intellect and truth be introduced. It is of paramount importance to the moral development of a child that from early infancy up to the age of twelve roughly, he is able to look up to the adults in his environment and see them as examples to justify his enthusiasm and engender a desire to emulate them. Parents and teachers and, in fact, all adults should know that they are those who should inspire in a young child not only feelings of love, but equally feelings of admiration, respect and veneration, without which human existence remains impoverished.

In his early years, a child's critical faculties and intellectual independence should not be forced forward, so that when he arrives at puberty, the point at which these faculties start to develop naturally, his moral codes will already have been properly constituted and consolidated, and can safely withstand the erosion which naturally tends to come from his own developing critical intelligence.

One of the most disastrous and dangerous tendencies in modern education is the forcing of intellectual and critical independence in children when they are too young and ill-equipped for it, only to have authority reimposed on the resultant unruly adolescent. This is precisely the wrong way to go about things and can be the source of serious damage to a child as well as "on the bounce" to society. As has been indicated, a solid foundation in the ability to respect and to venerate must be laid in early childhood to allow for critical independence later on, which will then not be destructive and at variance with the young person's moral development, but which will equip him far more fruitfully for the role he will later play in society.

In adolescence, the developmental needs of individual handicapped children somewhat diverge. There are those who benefit from continued schooling up to the ages of seventeen or even eighteen. Others require early introduction into practical work or craft training, if maturition is to be promoted, and what to do in each single case must be decided on its separate merits.

As a first example, craft training can be used therapeutically. For example, a child whose hearing and listening ability is impaired owing to faulty movement co-ordination, as was described in an earlier chapter, will benefit from training in pottery, because the high degree of co-ordination necessary to mould spinning clay effectively is one of the best exercises in guided co-ordination, which is the basis of the ability to listen and to hear.

Likewise, children with defects in their visual motor-co-ordination benefit from being trained in weaving where particularly this ability must be intensely applied. An over-sensitive and hesitant large-headed child will do well to learn gardening, which will draw activity from his head into his limbs, and the heavy-limbed, small-headed child will derive similar benefit from learning crafts that require finer application.

However, a moment comes in the development of a handicapped child when a complete change in attitude is indicated, when it is no longer fruitful or even permissible to use work and training in crafts therapeutically, and when a person must be regarded as having completed his childhood and as being on the way to his own adulthood. By this time, handicaps, limitations and odd inclinations, pathological as they may seem, must be accepted as part of a person's make-up, and will now have to be integrated into the wider community.

The choice of training and work take these factors into positive consideration, so that the fullest use can be made of the peculiarities presented by the individual's constitution, for it is essential that these peculiarities do not exercise him or set him aside.

The first step in the new, non-therapeutic phase of training should be less predominantly geared to the learning of one particular craft or kind of work, but to the implications of a full working day. A youngster should experience that he is no longer learning for his own benefit and self-development, but that he is setting out to learn to work for the needs of others. This new element in motivation should not be supported by the spurious idea of reward, but by the infinitely more genuine idea of the satisfaction derived from work that is well done. Hence the importance of choice of work suited to the handicaps, limitations and predelictions of the particular handicapped young person.

It is most gratifying to witness the eagerness, joy and determination with which handicapped youngsters take to this approach to work, particularly when their earlier schooling and upbringing was based on developmental insight.

Naturally, the time each individual takes to establish himself in the working day differs. Some may achieve the necessary perseverance and continuity within weeks. Others may take a number of years before they are sufficiently mature to sustain a working day. By and large, it is not before the nineteenth or twentieth year that a full working day can be sustained by the majority of handicapped adolescents, and they must be allowed a number of years for maturition to be achieved.

The other aspect of remedial education should now be taken up, which is the learning and acquisition of knowledge, skills and abilities. A principle of modern, scientifically conceived, education is that learning can only be individual and that it must proceed in steps small enough for the child to go from success to success. In remedial education, we must add to the smallness of steps a qualitative differentiation that is based on an individual child's developmental handicap.

It is obvious that for remedial teaching of specific skills, the principle of classes based on the same chronological age, so helpful for the child's development as a personality, does not apply, and that resort must be taken to individual teaching according to individual need.

The remedial teacher must always take two basic steps into

consideration—first of all, he will have to accept tactfully the child's handicap and limitations, but then lead the child resolutely towards the conquest of its limitations. This may sound simple, but all too often the first step is overlooked and the conquest of limitation as such alone is striven for.

For example, there is a large-headed child's short span of attention and his preference for abstraction and symbolisation. These must be taken as the starting-point in teaching and only gradually should he be guided away from his abstractions to the more emotional participation in things and from there to actual activity, which may consist of drawing, painting or specific limb-activities.

On the other hand, as a further example, there is a small-headed child's need for repetitive action, of pure *doing* rather than *thinking* about what he is doing. He will, for instance, "write" with great energy before he has any notion of the significance of the letters. Here, activity must be carefully and gradually led over into, or reduced to, the ability to symbolise. His starting-point is completely opposite to that of his large-headed peer.

Therapeutic approaches to different developmental handicaps described in previous chapters will indicate the necessary teaching approaches, the over-all principle being the one I have mentioned above, i.e. the acknowledgment of the child's specific peculiarities and handicap as a basic attitude. Again, this sounds simple, but there is the misconception that this is what the child really wants and that to give in means to collude with the child's frailty. But particularly when one is witness to the pathologically restless child's wild tearing-about and attacks of self-aggression, one cannot say that the child *wants* to be like that. He cannot help being like that, but he will be grateful when his condition is acknowledged and will begin to respond to the compassionate efforts to release him from his suffering and to lead him to learn to control his hyper-activity.

In an atmosphere of compassionate but also of compelling warmth, the most severely handicapped child has every chance of achieving the triumph of humanity over his affliction.

11. The Mongol Child

In the first chapter of this book, there is a brief reference to the difference between the medical and the developmental interpretation of mongolism. The general phenomenon of mongolism, the appearance of the mongol child in our midst, is now taken up anew here.

The mongol baby is a baby like other babies, only much more so. He is rounder. His head is rounder, though a little flattened at the back. His face is more circular, and a little dish-like in appearance. His eyes and mouth are rounder, and his whole body is softer than that of other babies. (For those who love babies, the mongol baby is an exceptional source and object of love.)

As was said earlier on, this is the appearance of every embryo at about the second month of pregnancy. Another two months earlier, the embryo is, so to speak, completely round, a minute sphere, a fertilised ovum, an All, but an All only of promise, of not-yet-unfolded future.

In various ways and to various degrees, the mongol child retains certain of these early elements of roundness, of undifferentiated promise throughout his life.

All his developmental milestones are delayed. He is late in reaching out with his gaze and with his hands. He is late in sitting up, in standing and walking, he is late in developing speech. There is often the danger of his being *too* late, of only learning to walk at the time when he should have developed full speech and, therefore, of it becoming too late for him to do so. By and large, however, speech is achieved and the humanity of his existence reaches at least that degree of fulfilment afforded by walking and speaking.

The mongol child usually remains half-way between the widely expanded consciousness of the infant and the centred consciousness of the mature adult. He practically never acquires that degree of separateness or alienation from the world that characterises the ordinary man of today. The mongol child remains uniquely at one with his surrounding and particularly with those people more familiar to him. It is as though he experiences everyone as brother and sister, or as fathers and mothers or members of his family if they belong to an older generation. He retains a family-feeling for the whole of mankind and the experience of the "stranger" is strange to him.

His powers of imitation continue to accompany him, uninhibited by self-consciousness, so that not only does he live *with*, but he practically lives *in* others. He has a great sense of occasion and of the theatrical, and he is usually not only himself and self-contained, but he is everybody and everything else as well.

The mongol child is given up to sympathy because the doubt that is engendered by knowledge has not yet stung him. His love is not that love which is born of pain and consciousness; it is, in a sense, original love, love that is innocent and entirely unburdened by intellectuality. It is almost as though "his eyes were not yet opened", as if the aura of human existence before the Fall from Paradise were still about him.

Yet—he can feel shame. Shame seems to be a deeply pervading emotion in a mongol child. While on the whole he has little sense of fear, he can be deeply gripped by a shame which engulfs him and prevents him from achieving things that he would be well able to do.

This experience of shame in a mongol is not the shame of the body or of its nakedness—it is a much more archetypal kind of shame, the shame of the nakedness of birth, of being born with an insufficient clothing-of-body.

An ordinary person, an adult more so than a child, protects himself by his faculty of dissimulated deceit. A mongol child in his original, pristine, unspoilt naivety lacks this self-protection. The mongol child is exposed to the scrutiny of life in a way experienced by no other handicapped child, however severe his condition. This being born insufficiently "clothed" by the body is what sets him apart on the one hand, and what calls forth the great amount of sympathy, protection and compassion from those who know him, on the other.

The peculiar combination of the abundance of love in the mongol child and his lack of intellectuality is both his charm and attraction as well as his frailty and undoing, for it renders him nearly incapable of maintaining himself on his own in our present day life and society.

To this should be added the fact that the mongol person hardly ever develops sexual activity and reproduction. Only very few cases of mongols becoming parents are known and even these seem uncertain.

What significance can possibly be discerned in the phenomenon of the increasing numbers of mongol children in our midst over the past hundred and twenty years? It was not before 1850 that mongol children were first described and then, interestingly enough, they were described as being a deviation from another type of congenital handicap—cretinism. Cretinism is a form of developmental handicap as old as human history. Cretins were well described in antiquity, and it is likely that some of the fools in Shakespeare's dramas are characterisations of cretinoid persons.

The nature of the cretin is polar to that of the mongol. The cretin is a hardened, cunning person. He matures quickly, possesses sharp wit and intellect, and is astonishingly and exceptionally prolific in procreation. In the earlier part of this century, large cretin families could still be found in mountain valleys of the Swiss Alps, families sometimes containing eighteen to twenty children.

Cretinism, however, has largely disappeared in the course of this century as a result of the discovery that it is due to a malfunction of the thyroid gland which can be prevented by the addition of iodine to cooking salt.

To return to the mongol child: people behave to-day as though sexuality and intellect were achievements of this century. At every street corner, in every advertisement, every cinema, every theatre, every journal, we are confronted with some aspect of sexuality and/or intellect. Sex and intellect are taking possession and have moved into our schools. They occupy an unprecedented position in society.

Then, in all this, the mongol child appears—loving, innocent, unintellectual, helpless, and so very appealing. Perhaps the cretin has accompanied mankind from the dawn of history as a kind of forerunner to our time, as a form of premature sexuality and intellect, and, having fulfilled his mission like Shakespeare's fools, can now vanish? And has not the mongol in some way mysteriously taken over to provide a medicine rather than to be an illness in our time? Does he not signify something we ought to learn to understand, to accept, and to love for the sake of our own development as Man?

Such thoughts may sound strange in our modern age, but still stranger is a mongol child's being, as an echo, a reminiscence of the Buddha, so much so that a mongol can hardly help but sit with his legs folded under him in the Buddha-position, as do those who are not out to change the world by main force, but to be gentle and to change themselves.

In the frequent and intimate encounter with a mongol person, it can dawn on one that he is, perhaps, a kind of messenger come to remind his fellows that, in all their technical advancement in the pursuit of original causes, and in all their power to change their environment, they also have a mission to change themselves.

Only a few hundred years ago, men regarded the limitations of their physical environment as absolute and insurmountable and applied their urge to develop and progress rather to the ennoblement of the mind. Equally, not long ago, it was established knowledge that a body heavier than air could not fly or that nothing of mass and volume could move through the air faster than sound, or that matter could not be divided into components smaller than the atom.

Technical advance began with the realisation that these given physical limitations can be overcome, something which our own century is clearly demonstrating. If, however, the overcoming of human egoism is mentioned, the wise shake their heads, knowing that egoism is inherent in man and cannot be changed. Yet our violent and aggressive century has also shown a degree of compassion, tolerance and social responsibility that is new in man's history.

There are distinct phases in the evolution of humanity and our own time is certainly no exception. Perhaps the mongol child is a messenger from a distant future in which the values men have will differ radically from those of the present?

Scientific advance will make it possible soon to prevent the birth of mongol children by early detection of aberration in the genetic structure. Will we gain by this? Or will we perhaps lose by it? Are we perhaps trying to do away with the medicine before we have derived the full benefit from it?

When such thoughts assume the proportions of experience, one encounters the mongol child differently. One does not see only the pathology of his condition. In him one meets a new brother.

A mongol child, then, does not require any special education or therapy. He should be allowed to share in the provisions made for other children, not only to be helped himself but to help others. Ideally, if there were one mongol child in each class in all ordinary schools, acknowledged in his human dignity by the teachers, the other children would be helped to develop forces of empathy and love which would later lead to new social advancement in society. More immediately, it could effect valuable changes in our educational system.

Nonetheless, a word can be said about the relative academic achievements of mongol children. These can vary considerably. There are some mongol children who learn to read and write at a fairly early age and who will later read books with full comprehension. Not a few develop an interest in history. On the whole, they tend to be "humanistically" and not "scientifically" minded, although some mongol children are known today to take to arithmetic, in spite of the prevailing view that mongols cannot reckon with numbers. There are other mongol children who never learn to read or write, but whose general skills can be developed sufficiently to allow them to take responsibilities in different fields of work later on, even in the domestic field which is one that requires more foresight and planning than the routine assembly work in many factories.

Equally, if mongol children share in the life, education and training of children suffering from other handicaps, they prove to be invaluable healers and helpers of others. They call forth, love, empathy and responsibility in children who otherwise have the greatest difficulty to develop just these qualities. They manage as no other person does to overcome the isolation and withdrawal of the autistic child in their blissful awareness of the lack of response from them. Their ever-forgiving and utterly unreasonable degree of sympathy can provide the severely maladjusted child with his first experience of being wanted.

All this is not intended as a glorification and apotheosis of the mongol child. I do not mean that I want the mongol child to be a mongol, but that we shall learn to live with him and accept him as one of us, different and yet essentially, deeply, utterly human—as, in fact, our brother.

Under circumstances born of the right attitude, the mongol child will grow up into astonishing maturity and responsibility. He will gradually develop conscience and understanding in spiritual matters. For, if, to begin with, the mongol child seems to lack religious feeling, if he seems to have no conscience and is only full of fun, this is because he cannot encounter evil as other people do, either in himself or in the world. He is truly innocent.

Yet his innocence must be rightly guided. If he is allowed to continue to regard life as a game, he cannot integrate as an adult into the world of adulthood. If he is accepted and taken seriously for what he really is, he will mature, become responsible and reliable, a valuable friend to the community in which he lives.

IV

The Environment of the Handicapped Child

So FAR, childhood handicap has been described from the point of view of individual development, which is a child's effort to integrate his Self into its inherited bodily constitution and to express his personality by means of this constitution.

Handicap, however, only comes about in relation to an environment. A child is not "handicapped" within his own scope of existence; but only in the situation into which he is placed. To begin with, there are the standards, customs, and values set by a particular environment in comparison with which an individual is "handicapped".

Having introduced the study of a variety of developmental handicaps by a brief survey of child development, it is appropriate to engage in a brief contemplation of the part "we ourselves" as part of the child's "environment", play in the occasioning of his handicap.

In order to do this, we will have to examine ourselves in our encounter with children, whether we be parents, teachers or simply other persons. Here, a very wide and rich field is entered which will, by no means, be exhausted in this chapter. We may, however, glean some few grains of understanding which will enable us to form, fortify and rectify attitudes in a way that will give support to the further development of a handicapped child.

The earliest environment of the child—the family—provides the obvious starting point, then the handicapped child in school, and finally, the handicapped adult in society may be considered.

1. The Family

When a child is born, the mother anxiously asks: "Is my child all right?" This question alone contains several aspects. Firstly, has the child come unscathed through the dramatic and possibly traumatic event of birth through which he and she have gone together? We ask a similar question when someone we love has undergone an operation. Like birth, surgery is a dramatic event which may lead to a positive end and assure life, but it is also fraught with risks and dangers.

But the mother's first anxious question contains something else. "Is my child all right?" also means "Is he physically perfect? Are there no malformations? Has he all his limbs? All his senses?"

Ultimately the question, "Is my child all right?" means "Will he live? Will he grow up? Will he become a man? Will he not be handicapped?"

The way in which this first anxious question is answered is of fundamental importance, because an unfortunate combination of elements of guilt and fear can inflict irreparable hurt on the harmonious interplay between the child and his environment.

Feelings of guilt are often experienced by the parents and family. Fear and uncertainty, however, tend to arise more in the medical and professional situation or environment surrounding the birth, something which will be returned to later on.

Feelings of guilt may be the result of ideas about heredity and the genetic make-up, which, of course, do play a fundamental part in the physical make-up of any child, but it must be realised that genetic make-up from a present-day viewpoint is shown to be so vastly complex a problem that it is impossible for an individual to assume the responsibility for what he has inherited and for what he is passing on to his child as a genetic foundation. Hence parental feelings of guilt on this point are without grounds.

There is, however, a much more essential aspect to be considered and that is that though the child's constitution is determined by genetic factors, his *development* is not.

As the reader will remember, we said in the second chapter that development must needs go through frustrations, and that the furtherance of development does not lie in *avoiding* frustrations but in making frustrations *meaningful*. In this sense, the genetic make-up inherited by the individual must be seen as one of the basic "frustrations" that must be accepted from the outset, and made the best of in endowing it with meaning, thus creating a personal biography.

It must also be seen that constitutional frailty must become more frequent with increase in survival due to the advancement of hygiene and medicine and must therefore be seen as one of the outcomes of human civilisation.

It should also be realised that even among the relatively few genetically determined developmental handicaps, still fewer are genuinely hereditary. Mongolism, which is now known to be of genetic origin, is only rarely of a hereditary nature, the great majority of cases of mongolism being due to as yet unclear changes in the genetic make-up.

Feelings of guilt can also be attached to actual pregnancy and birth. The mother may think that things she has done or omitted to do during pregnancy, or personal situations at birth, may have caused handicap in her child. As a rule, however, such feelings of guilt are more often than not absent, because most mothers still sense, at least instinctively, that during pregnancy there is not that degree of separation between them and the embryo that would allow for responsible influence on the developing child to take place. The oneness of mother and child in embryonic development, as well as the genetic situation in which mother and child are indistinguishable, is borne out in an interesting way by old folklore. For instance, the fact that a child is born with a hare-lip is ascribed to the mother's having been frightened by a hare in the early months of pregnancy. This is the peasants's way of instinctively describing the following.

When a mother carrying a child whose laterality-development in the early months of pregnancy is such that the lips fail to fuse completely between the left and the right half, is confronted by a hare, she suddenly recognises in the hare something of which she is as yet completely unconscious, but something that is actually happening in her, as part of her, to the developing embryo. Fanciful though folklore may seem, one only need listen to the experiences of hundreds of mothers to realise how real they are, and how they are an expression of the *oneness* between mother and child before birth.

Really powerful and profoundly pervading feelings of guilt, however, are not the result of superficial situations or interpretations, but rather the work of the subconscious layers of the mind or psyche.

In his emotional make-up everyone has certain unresolved problems, and those states in which one is in basic disagreement with oneself are easily projected as feelings of guilt into the situation of the handicapped or malformed child.

Developmental failures in a child can be equally subconsciously attributed to inner conflicts not consciously faced, and this can mar and burden the early relationship between mother and child as well as between family and child.

Although the underlying problems may not be completely resolved, it is often possible to clear the relationship of parents to their child and ease the burden on both, if the mother and the rest of the family can be made to understand the child from the developmental point of view.

The first task is for the parents to learn to see their child essentially as a promise, as a potential. The child is not merely what it presents to view; it is equally what it is to *become* in the future.

With all his potential, the child, as soon as he is born, is part of the family; he is not something separate. He belongs to his mother, his father, his family and with them, to the totality of mankind, sharing in the potential perfection of mankind as well as in all its obvious imperfections, which manifest in the stream of heredity generally, and more specifically in the family situation.

Every child goes through the phase of complete dependence on and oneness with his family, which is his intimate early environment. Nonetheless, he is on the way to *himself*, to becoming himself distinct from his family.

This may seem highly platitudinous, but it is precisely here that so much initial support can be given or so much initial frustration lies. If parents realise their oneness with their child as a family on one hand, and the potential uniqueness of their child on the other, obstacles will not be put in the way of their child's development. If there is confusion in the family concerning these quite distinct elements, which is frequently the case in families with handicapped children, the family itself can prove to be the initial obstacle.

That clarity be established as to these things in the minds of parents is an essential requirement.

Returning to the element of fear in the birth situation of the child, it was said earlier, that this is more often evoked by medical or professional aspects of the environment. It is in fact not restricted to the mother's question about her child immediately after birth, but holds good for any situation in which parents seek advice about their child from experts, especially from doctors.

The medical profession is trained to cure or at least to alleviate disease, to ease pain and prolong life and tremendous advances have taken place in this field in the past few decades. When the medical profession is now faced with a subnormal child for whose condition no cure or treatment is known, the result is helplessness and hopelessness. The consultation ends on a note of finality, and this sense of impotent finality gives rise to fear in the parent. In spite of the compassionate attitude of most physicians and social workers, it is the note of hopelessness that communicates itself to the parents of a handicapped child.

This, unfortunately, is a frequently occurring situation, because parents naturally refer to the medical profession with the question "Can something be done?" which the medical profession can only answer in its own terms. Thus there is collusion and misunderstanding on both sides, and the misunderstanding concerns the fact that it is not a question of "cure", but of the development of a handicapped child which is to be witnessed and helped, not by changing the child or the condition, but by changing the understanding and the attitude of persons in the environment of the child. Just here lies the meaning and urgent necessity of seeing child handicap from a developmental point of view.

A handicapped newcomer in the family may, of course, present a problem to his siblings. It is likely that he will absorb a great part of his mother's time and attention. Some handicapped children mature so slowly that it is almost as if the mother were having a series of babies for a number of years when merely looking after the one. This can prove strenuous for all concerned and has to be dealt with as best as it can within the family.

There is another problem which results from the different standards according to which normal and handicapped children have to be dealt with, which also exists when there are children of different ages in one family. The attitude of normal children in a family depends entirely and exclusively on the attitude of the parents and on their degree of mistaken attribution which blurs the true picture of the proper course of events. If parents have the sound, clear attitude described above, they will find no more eager and co-operative helpers than their own normal children. But if parents feel that it is unjust to apply different standards to the handicapped child in the family from those they apply to their normal children, the latter will naturally feel resentment.

Young children up to the age of seven or so participate to a marked degree in the conflicts and experiences that go on inside their parents without a need for overt communication. When parents learn to manage the ambivalent relationship between themselves and their handicapped child, learn to manage both the symbiotic aspect in which their child's needs are their own needs, as well as the individual aspect in which their child's efforts are experienced as those of a separate and free individuality struggling to establish it, the other children in the family will not only show that they are willing to help, they will be masters at the game. They will not have the slightest difficulty in respecting the particular needs of their handicapped sibling and at the same time, in treating him as an equal.

Unresolved, subconscious conflicts in the parents in respect of basically accepting their handicapped child can cause severe damage to the other normal siblings even if an overt positive decision has been made on this account. This will not so often result in aggressive reaction from them, but rather in withdrawal, regression, severe nervousness and possibly real emotional breakdown, particularly if the normal sibling is frail of constitution. Such situations, unfortunately, lead more frequently to the exclusion of the handicapped child from the family than to the solution of the parents' basic problems.

However, as has been said earlier on in another chapter, it is essential that the entire family, though more particularly the parents, do not surrender their own needs and standards of living to the needs and abnormal standards of living of their handicapped child. Not only does the family possess the means to frustrate the child, the child possesses equal ability to frustrate the whole family, which in turn only serves to aggravate his lack of integration.

There must be a symbiotic relationship in which the requirements of all members are respected and brought into harmony in the best possible way.

This, of course, is more difficult when a handicapped child is an only child, for then the parents are tempted to arrange their lives exclusively according to the needs of the child, something which, if not handled very carefully and circumspectly, creates an environmental situation which petrifies the child's problem.

The specific needs of a handicapped child must, therefore, always be dealt with in context, and it is inevitable that there will be sacrifice on both sides. When sacrifices, however, are regarded as frustrations that are meaningful, both family and handicapped child will have achieved a fundamental step in mutual development.

Just as prolonged infancy in a handicapped child imposes a strain on his mother, so does the frequently prolonged toddler-phase, the self-willed, undisciplined, destructive phase of early childhood, tax the patience and power of endurance of the whole family, particularly when it lasts right up into school age.

A wise and mature way of dealing with a child will have to be worked out by the whole family, a way which at no time leaves him in doubt of the family's, and in particular, the parents' love for him. Even if he causes hurt or has to be restrained, this love must be sustained. Punishment, which is almost always prompted by a sense of retaliation or revenge, is under no circumstances adequate as an educational method.

If it is correct to maintain that the central argument in the education of the young, pre-school age child is the *good*, it follows that the child must learn to experience that certain things he does are hurtful to the other members of his family. It is essential, therefore, that the parents react because they are hurt, anxious or frightened, and not because of some principle of justice or education. A child needs to experience the joys, but also the pains, disappointments and injuries which his actions cause, but not at second-hand. He must also experience the reprobation with which his family regard some of the things he does, even though he is not cast out of their lives for doing them.

A handicapped child must be able to feel that the whole family is unshakably convinced of his goodwill, convinced that in truth, he wants to be loving and to do what gives them pleasure. When he falls short and does quite the opposite, the family does not construe this as a premeditated, deliberate and malicious intention, but sees it as a lapse, a failure, which the child himself would dearly like to avoid the next time. The family and particularly the parents, must bear in mind that success or lack of success in refraining from or achieving action is no true measure of intention. However often a child may fail, even if he always fails, his intentions may be positive, and only their realisation delayed.

This atmosphere that "keeps no score of wrongs" is the necessary one for any child to grow up in, all the more for a handicapped child. It can best be provided for the young pre-school child in his own family, and therefore, whenever possible, he should in any case remain with the family up to school age.

2. The School as Environment

In many cases, the needs of more severely handicapped and disturbed children cannot be met adequately in day-school. This is because the family of a school-aged handicapped child is often unable to deal with the situation of having the child with its growing needs in its midst. Day-school will provide education as well as hours of relief for both family and child, but the basic tensions to which a child returns each afternoon will ultimately prove to be a negative and preventive influence on a child's developmental progress at school. Therefore, in such cases, admission to special residential schools is indicated.

Residential or boarding-schools have for a long time been a characteristic of the British educational system for normal healthy children. The set-up of the boarding-school is based on the normal resilience of a healthy child and his ability to adapt to the distinct and differentiated systems of frustrations that boarding-school life presents, in the knowledge that, relatively cruel though the system may be, it has strong character-forming and hence meaningful features.

A special residential or boarding-school for handicapped children differs fundamentally from a boarding-school for normal children. It need not—indeed it must not—create a system of frustrations to be met by resilience and powers of adaptation, because each handicapped child arrives at school with his own existential complex of frustrations, and will, by and large, appeal to the *caritas* and the protective attitude of the staff of the school.

The *caritas* and protective attitude on the part of the staff of special boarding-schools tend to be the more pronounced, because of the fact that a young handicapped child leaving his family for the first time in order to go to school is in all probability still going through an earlier phase in his development in which, by rights, he is still in need of family care and guidance, however wanting these may be in actuality. For this reason, a young handicapped child arriving at boarding-school is especially vulnerable.

As a natural reaction to this, the staff or of some of its members will want to re-create a family-situation for a child within the school. This is frequently expressed by the fact that a member of staff becomes "mother" or "father" to a child and acts in *loco parentis*. While this is a possible course to take in a foster home or in cases of adoption, it is inappropriate in a school, and when it is pursued, it gives rise to new problems, for example, tensions between the real and the acting parents, and between the latter and the rest of the staff.

It is altogether questionable whether a parent-to-child relationship can be imitated, re-created or substituted for, because, other than the real parents, no one is participant in the unique biological-emotional symbiosis of parent-and-child, even where actual home-circumstances are unsatisfactory. Even when a child is deprived of his parents at an early age, the biological factor and emotional experience peculiar to the family cannot be transferred to a person who is biologically "other".

Nonetheless, in a child's relationship to his parents there are powerful elements that transcend its biological-emotional texture, and which can be taken into consideration. In the senses of Jung's *anima* and Freud's *imago*, mother and father represent far more than merely themselves as persons. Through their convictions, beliefs, moral standards and values, and just as much through the absence of these, the child has his initial experience of the world. What these convictions, beliefs and moral codes relate to is less essential to him than the strength with which they are upheld. From this strength of conviction a child will derive his basic "religious" experience. Also even a child whose parents are confirmed atheists will have a *religious* experience of trust and faith through the firmness of his parents' convictions. To a young child, his parents are the representatives both of the world he does not yet know, and of ultimate, spiritual mother-and-fatherhood. Bound up with these experiences, he has his primary sense of security and meaningfulness or its alternative, annihilation and despair.

A personal relationship of a child to his parents does not generally emerge before the onset of puberty, but after this, a young person will begin to see his parents as they are and, according to circumstances, he will cope with their frailties and idiosyncracies. But he himself will no longer need them to uphold his experience of primary meanings, values and beliefs. These must now grow out of himself and be stimulated by other influences. Yet, in spite of his "emancipation", the original biological-emotional relationship to his parents remains the cornerstone of his existence.

This, too, is the case for deprived children, because the primary need is not obliterated by the loss of the parental context. Here, everything should be done to help a deprived child to accept its given circumstances which are those, one might say, of biological-emotional surgery, and at the same time, help him to feel that even this might be experienced as meaningful, in that the loss of something is not only a deprivation but also a challenge to be met in a constructive way.

To return to the young handicapped child who must leave home at an early age and enter boarding-school because he is in need of that education and therapy which a special school affords—and who may look forward to several years away from home. What of his need to experience-the world through his parents? What about his feeling of primary security?

A handicapped child is both more vulnerable and less articulate than his normal brothers and sisters. Not even at home where he is embedded in his family context can he voice to his parents or to himself his experience of primary meanings, which does not mean of course that he does not have these experiences. But they are diffuse, nameless and without form and shape. Hence, all the more does a handicapped child require an articulated and differentiated school-environment in which to grow up and to learn.

To provide this articulated and differentiated environment then, in which a handicapped child can himself become articulated and differentiated as a personality is the task of the school as the child cannot remain at home.

How can this be done? Can it be done at all?

I believe, with reason, that it can and will be done in a variety of ways, and that the way of "community" is the way to take.

I have had experience of a residential school run as a community for thirty years and would like to submit a description of this community, as it is one attempt which, I hope, has been accompanied and which will be followed by many others.

At the outset, it should be said that the community of which I am going to write has learned a very essential lesson. It is that you can have a community *for* handicapped children, but if you have a community *with* handicapped children, it will be a fuller, more effective community.

This principle is, of course, realised in some psychiatric hospitals today where the "therapeutic community" is made up of everyone who has something to do in the hospital, and the principle is applied very specifically if some of these persons happen to be patients.

There are two essential functions of handicapped children in a school run as a community. The relationships between adults in a community are in some ways analogous to marriage. In marriage there must be a constant attuning to and retuning of mutual relationship and frequently, the children of the marriage provide a dimension for the relationship that nothing else can provide. In like manner the children in a community of the kind we are going to discuss add a dimension which leads out of and beyond the introspective adult world into the future—for no one can deny that the future lies with our children, even though they may be handicapped.

Finally, one of the first conditions of a well-functioning and vital community is the encounter of the individual with himself, is the facing of himself and the knowledge of himself that results. Every individual can find elements of his own existence in the general pathology of human development. There is no one entirely without those elements such as described in this book as being handicaps and disturbances in childhood development, but deviations and one-sided tendencies are for most of us contained sufficiently within the bounds of normality for the "normal" person to achieve self-expression. The mirror of human developmental pathology is so clear, so unadulterated in its reflection of handicapped children, that the person gazing into it sees more of his own and his fellows' fundamentally human significance, more of its vulnerability and the danger, and much more of the meaning of human existence altogether, than he can see by regarding himself in the mirror of "normal" environment.

I do not mean "There but for the grace of God go I!" I mean instead—*this* child has something to teach me about the grace of God which is none other than the most primary of meanings. This is the reciprocal bond, the give-and-take between children and the adults to whom they are entrusted for their education and training. They come to take what their teachers can give and in return, they give their teachers self-knowledge with a tender, generous but unmistakable gesture.

Much thought and more experience have taught the particular community of which I am writing that a certain differentiation amongst spheres of life has to be recognised and each sphere given its own expression. For the purpose of characterisation, one might say that Body, Soul and Spirit will serve as a kind of frame on which to construct these spheres: "Body" pertains to the *work* of a community with its *economic* implications: "Soul" pertains to the *living-together* of a variety of people in a community, and "Spirit" pertains to the *idea* or *ideal* which is the transcending and encompassing site of unification without which, I believe, a community is not fully a community, and is at best only a partial community. There can be, and there are, communities in any one of these three spheres. There are economic associations, therapeutic communities or religious orders, which may be quoted for the sake of illustration. But another interpretation of community would seem to necessitate the conscious inclusion or exercise of all three spheres: work and its economic implications, the mutual human encounter of those who do the work, and the common idea—or one can say if one likes, the common religious or philosophical background, or ideal.

Differentiation and articulation in these basic spheres of human existence in community lay a foundation for that coherent and articulate environment we spoke of earlier as essential to a handicapped child's way-of-experiencing the world in the pedagogical situation as distinguished from the biologically and emotionally determined family situation, and it ultimately constitutes what is known as "care" in a residential environment.

One of the first things that emerges is the application of that principle I have characterised as "Spirit", or in other words, religion or philosophy. Must a community for handicapped children under all circumstances be a religious community?

I would say that in the Western World, it can and should be a "Christian" community, with which I do not mean necessarily "Christian" in the sense of denomination, creed, or orthodoxy, but Christian in so far as it is the cultural and spiritual heritage of the last two thousand years in the West to have partially developed and held high such values as those propounded by Christ and passed on through the Gospels. Christian, too, in the sense of the search for what Toynbee calls "the God of Love" or "the ultimate reality".

These are existential matters today and are issues upon which eventually our Western Civilisation will hinge as western technical science advances.

I stated earlier that I believe it matters less to a child *what* his parents believe than the strength and fervour with which they believe it, for the latter will inevitably create their way-of-being and hence their child's way-of-being in the world.

Likewise, in a pedagogical community, the particular creed or philosophy may be a debatable point. Nevertheless the certainty of conviction and firmness of belief of that community will provide a spiritual discipline as well as a mantle of security and meaning. It is only when meaning is lost that insecurity results. The world without meaning is a dangerous and threatening place. Spiritual discipline is an ingredient of the bulwark against meaninglessness. But it is only an ingredient. There must also be the continued search for what is of primary significance in human existence and in a community, expression must be given to this search.

The community of which I have had experience attempts to do this in various ways and by various means. The day starts with what is called Morning Prayer, although the words used would not refer so much to a Deity but would rather challenge the minds of children and adults alike to face the coming day with their best powers so that it is not simply "another" day, but rather a new opportunity, a singular event, for in truth, *this* particular day was never there before. Likewise, a grace is spoken before meals to create a moment of mindfulness of a greater significance in eating than merely to satisfy hunger and take in calories. Evening Prayer is said when the day has drawn to its end, again so that the night is experienced not merely as an extinguishing of consciousness, but as a dimension which transcends the day with its own peculiar restoring powers.

Before sleep, there is still the need of an individual child for "dialogue" with *his* God. In most children this will take the form that his particular family denomination has taught him. Others need a new expression of their innermost thoughts. When a parent dies, this "dialogue" with the personal God becomes an urgent need and space for it must be provided.

These daily practices culminate in the Sunday Services which are non-denominational forms of Christian service, and which differ for young children, older children and adolescents. In the response given by the children in the Service, a child has a moment of the unique presence of his own potential. This is manifest in that, although virtually all children attend the Services regardless of handicap or disturbance, the outside observer or visitor sees hardly any evidence of abnormality in the group, and sees rather a picture of great dignity and profound awareness.

The unique stillness of a large group of handicapped, psychotic, hyperkinetic and otherwise disturbed children is often put down to the over-forceful, possibly old-style non-permissiveness of the staff. I would suggest, however, that the obvious dignity and "at-one-withness" of children in these situations cannot satisfactorily be accounted for in this wise. On the contrary, the old proverb about leading a horse to the water but not being able to make him drink applies as much here as it does to a horse. That which makes a horse drink is its thirst. Thus a child can be conducted to a Sunday service in a disciplined way, but what ultimately makes him partake is his own inherent human need to experience the "ultimate reality".

On the day before the services, that is on Saturday, all children have their "Religion Lesson". The teachers who take these "lessons" attempt to be in no way doctrinaire, but rather try to discover the needs of a particular group of children and by means of imaginative material such as parables, legends and the like, to provide ideals with which the children can identify themselves and in so doing, fortify qualities such as trust, helpfulness, compassion, abhorrence of injustice and so on. Through the archetypal dramas told in the Old Testament, for instance, children experience the epic of human evolution of which they themselves are a part. Job is no mere tale of hardship!

These "religion lessons" become the place where children as they grow older and more aware of their own situation as handicapped persons, will discuss their anxieties and questions as to why they are as they are, as well as their fears of life and fears of death. The "teacher" cannot meet these questions by presenting a "system" or a dogma, but only by putting his own faith, however small, at the disposal of the children and serving rather as "midwife" in the socratic sense than as an expounder of truths.

Religion lessons with older children are often the spark that sets off the need for an individual child to have an individual talk with a teacher about his personal problems of insecurity, anxiety, of "failing" his parents, or of sex. Here, a teacher must be fearless but not ruthless, warm but not sentimental. He must in that moment embody all the conviction and firm outlook that otherwise the community tries to express in its forms.

The daily morning and evening prayers and the weekly services are embedded in the larger cycle of seasonal festivals, the expression of which plays a big part in the cultural life of the community. Naturally drawing on the traditions attached to Christmas and Easter, or Midsummer and Michaelmas, Whitsun, Advent and so on, the entire community of children and adults celebrates these festivals in the form of plays, pageants, musical presentations, festive meals, each festival demanding its own mode of expression. Some of these things are done by the adults for the children, others by the children for the adults, and many by children and adults together.

In the experience of the children while they are at school as well as later in retrospect, these festivals stand out as pillars in the course of time, as points of spiritual reference, and one might say, as milestones on their way to themselves.

As the years go by, more and more parents tend to want to participate in the seasonal celebrations of the community and will travel long distances to do so. Witnessing a Christmas festival in the large Assembly Hall in which the community has widened to include parents and other visitors, one experiences a unique merging of the two sides of parenthood: the biological-emotional, and the spiritual, expressed in the Parenthood of the Christmas Event.

In short, the "religious" life of the community aided by the arts is the spiritual matrix in which a child gradually finds his own significance and identity. He will grow up and return to the "world" and, within its own possibilities, he will have to find his feet in the religious sense, and many handicapped persons have outspoken and definite leanings in this sphere. His school experiences will have given him a sense of direction and coherence in his encounter with life.

A brief reference should be made to the position of the staff, in particular, the young, more short-term members. In a community, there will be orthodoxy and radicalism, the devout and the agnostic. There will be the searching and the spiritually unburdened, but running as a thread through all these diverse elements is a basic selfless urge not to do things for one's own advancement, but for the sake of others, in this case, for the sake of the children entrusted to the community and to whom the community extends its love. It is upon human *goodwill* that an attempt on the part of the community is based to express meanings of existence and it is through goodwill that this attempt can become effective.

Needless to say, there are opportunities for communication among the staff on these matters, the most established of which is the so-called weekly Bible Evening, where the various house-communities have a modest but festive evening meal which is centred round a reading out of the New Testament to be read in the children's Sunday Services the next day. There is an exchange of experience, questions, problems and challenges that are stimulated in the individual by the reading of a particular passage out of the Bible, and in the course of time, many people of different creeds have participated. These weekly Bible Evenings have so developed as to include the older children on occasions.

Having attempted to deal with the sphere of the "Spirit" or *ideal* in community, I should now like to try to deal with the sphere of the "Soul" or rather, living-together in community. Living-together in a community of the kind of which we have been speaking, has two sides. One side pertains to the living-together of children and adults and children and children. The other side pertains to the living-together of adults. The former is both easier as well as infinitely more varied, the latter is more difficult and more confined to the personal encounter both with oneself and with the other adult person. I shall describe the former first.

Such a school-community should not be so small as to be unable to include the full age-range of school children, say, from the fifth to the eighteenth year of age. It should also be able to take children representing the greatest variety and width of handicap and disturbance, because the greatest variety will depict the totality of human possibility, whereas a one-sided amassment of any single type of handicap or disturbance will inevitably *stress* the *handicap* and not the individual. The staff then has to deal with *mongolism* or *autism*, with any *ism* and not with a group of growing children. The children in these circumstances will be confronted on all sides with the multiplication of their own problem. The pattern for the autistic child will be autism and for the mongol child, mongolism and he will, with all good will, be trapped in his own problem.

The community, however, should not be so large that it is difficult for everyone, staff and children alike, to get to know one another. A ratio of one staff member to two children is an ideal one. More as well as less staff to children causes strain on either side.

In spite of the overall numbers in the community which may range from 100 to 400, living accommodation is more conducive to living-together if it is not arranged on an institutional basis, but it is also not wise to go in for very small units only, or for standard-sized "cottage" units. Experience has shown that, whereas very large units tend to institutionalise a place, very small units make for the other extreme which is the isolation of individuals and individual interests from those of the whole community.

Ideally, there should be one or another large house and a number of houses of differing sizes down to very small units for only a few children. There are children as well as adults who do better in larger groups and equally others who have an obvious need to be in very small groups and many who do best in groups of intermediate size. In addition, the need for a particular group can change in the individual at different phases of his life in the community or in different states of health, so that it is helpful to have a variety of possibilities.

Within any single unit, there is a wider scatter of ages as well as handicaps in the children's grouping, and at all times, both sexes. Children sleep in small dormitories of three to five beds, occasionally in single or double rooms, and the room of the dormitory-mother or father adjoins where possible. The staff consists of house-parents and younger staff, and a selection of teachers, therapists, doctors or gardeners will be attached to this group.

Staff and children have all their meals together and share in the necessary preparation and clearing up of meals, as well as in the cleaning and upkeep of the rest of the house and its immediate grounds. Everyone, including the Superintendent, is addressed by Christian name.

This is the "home" side of life in the community. In the morning, however, children go to school in separate school buildings, often so far apart that transport becomes necessary and a child gets to know the extent of the grounds of the community. Each child, without exception, attends a class of children of his own chronological age and spends his school hours with a very different group of boys and girls from those in his own living-unit, forming school friendships and experiencing with his classmates the variety of subjects, projects, visits to places of interest such as mills, quarries and factories in the district which belong to the activities of his class.

For his artistic lessons as well as group therapies, a child will meet yet again different children and also other adults and so extend the network of differentiated relationships throughout the community with the result that, when the entire community is gathered for large events in the Assembly Hall or goes out to town to concerts or plays, he is completely at ease and adjusted to the large group, because in the course of his activities, he will have got to know everyone, both big and small, in the community.

It is essential that an experience of living-together should not be an exclusive one pertaining to the confines of the community. Already at school, handicapped children should be guided carefully towards living-together in the widest sense, that means, in the wider community. Children are therefore encouraged to develop their own interests outside the community and an increasing number, even of the younger ones, attend Brownie and Scout Meetings in the district, where they are not only tolerated by the normal children outside, but greatly appreciated. The community is constantly on the search for means of communication through which it can extend itself, so to speak, into the wider community, as this is needful as a further step in the realisation of the principle of non-segregation.

The community has gradually learned to recognise and, I would say, give the fullest possible play to the unique help that one child can give to another. That which the mongol can give to the autistic and vice versa, or the mutual assistance afforded between the maladjusted and the physically disabled. This help is often more spontaneous and instinctive than adults are inclined to think and when it is given the opportunity to unfold, it becomes a creative and therapeutic force within the whole. It is not that mutual help is "preached" to the children, or exploited when all else fails. But it should not be overlooked and frustrated, because a child's spontaneous urge to take responsibility for another child is often the first sign of a personality that begins to recognise and assert itself.

In short, if the great organ of varying relationships is consciously played upon in all its higher and deeper tonal ranges and not allowed to be merely incidental, a child will be beneficially exposed to a fair degree of social experience while still at school. Just as in the blind child there can be remnants of sight and in the deaf child remnants of hearing which need to be stimulated, so one might say that in handicapped and disturbed children there are "remnants" of social ability, social concern and responsibility. When these are called forth and cultivated, a child stands a good chance of being a useful member in whatever section of the wider community he may later find himself.

In a community such as the one I have been writing about, there is an inevitable coming-and-going of students and short-term helpers as well as more specialised staff. It is important to consider the element of permanency, to see how and where it is best upheld and where it must be clearly defined in the experience of the children.

The element of permanency is no doubt represented by the house-parents of the unit in which a child lives, but not infallibly, because for one reason or another, a child might move from house to house so that while he has been accompanied by one house-parent in his early years at school, for example, a different house-parent will see him through his final years.

Permanency will also be experienced through the superintendent, the doctors, therapists, nurses and many others who might constitute "permanent staff". It will be least of all provided by the group-parents, for these are for the most part students and young people who come only for a limited period. This, however, is not seen as detrimental to the children but rather as a challenge to their adaptability, for which the continuity otherwise present in the community allows. It would be different if the permanent staff were not *permanent*.

A very favourable factor at work in temporary or impermanent group-mothers and fathers is that *possessive* relationships are not so easily built up between adult and child. There are cases where a group-mother, for instance, possesses and is possessed by a child, which entails a usurpation of the parent-child relationship and places a social obstacle in the community. A free, warm, reciprocal relationship between group and group-parent is the most beneficial. Even a less free and less warm relationship between group and adult can be a meaningful experience for both sides as long as it is contained within the warmth of the whole community.

In this community, the greatest element of permanency is represented by the class teacher who remains with his class from its first to its final year, according to the educational methods adopted by the community. This means that a child having started school at the age of six and leaving at the age of sixteen will have had the same teacher throughout ten years. This is naturally modified when children are admitted at a later age, but nonetheless, they are admitted into a class which has grown together with its teacher so that they themselves enter a closely woven texture of long-term relationship.

The class teacher is the person who is concerned with the schooling of the mind of his pupil, and the experiences of the mind in contrast to those, let us say, of the body, are those closest to the experience of one's own identity. This is often a long hard way to go for a handicapped or disturbed child, but it is of paramount importance that he go it. The fact that he has *one* permanent, well-known and trusted guide is infinitely valuable and beneficial.

Before coming to the question of living-together as staff, something should be said about administration and leadership. There is an obvious trend in the world at large to move from hierarchically structured forms of society to community participation and responsibility, although no doubt these trends are still in the first throes of labour.

In the more confined fields of psychiatry and mental hospitals, there are already notable moves in the direction of community administration, where the old father-figure of the Physician Superintendent has given way to group-leadership.

The move from hierarchy to community is one of the vital issues of today and is, I believe, profoundly related to the revolution which is taking place in the relationship of man to his God. In spite of two thousand years of Christian tradition, men of the twentieth century are only now beginning to realise and experience Christ's words— even if they are not professed Christians. "I and my Father are One", which means that man and his God are one. Formerly, the crowned head and the church dignitaries represented God in society. Now potentially every man in every section of society represents God, or in other words, the community of men represents God. Representation of the Omnipotent is no longer a selective matter.

These are lofty words to introduce the modest modes of administration in a tiny community, yet in the microcosm are contained the dynamics of the macrocosm.

In this particular community which I am describing, the policy-making body is the largest body which comprises all those who consider *themselves* to be responsible and active carriers of the community. There is, therefore, no clear demarcation between "senior" and "junior" staff, as a young student may feel very responsible for the duration of his stay, whereas an older member may not be in the position to sustain responsibility.

Within this larger body, there are committees for Children's Admission, Staff Admission and Placement, the Training Course, Cultural Activities, Finance, Visitors and Building, all of which are responsible and accountable to the Community Meeting. As a point of reference for the world outside there is a Superintendent who, for the same reason, has his deputies, but he himself is not a member of any of the committees, although he has access to and can be consulted by any of them.

These committees are given the mandate to act on behalf of the whole in their own fields and this mandate must be constantly renewed through the confidence and co-operation of all the others. Each committee must endeavour not to transcend its own defined scope and not to usurp the authority of the whole community, and matters that go beyond mere arrangements and routine issues are referred to the larger body. Decisions are not taken by vote or consideration of the majority of opinions but by consensus. In a community, opinions matter less than goodwill. For the individual member it matters far more if he is able to decide for a certain move with all his goodwill, even if the proposed move seems alien to his opinion. If he withholds his goodwill, and sticks to his opinion, he might have lost the opportunity to learn a necessary lesson.

There are further ramifications of the administration, such as the single house-communities, each of which administers its affairs in conjunction with the whole.

Then there are three "colleges", i.e. the Teachers' College, responsible not only for arrangements in the field of schooling but also for the continued vitality of the education in the school; the Child Guidance College responsible for the development of the child as a *person* and his maturition and social integration, and finally; the Therapy College, responsible for all the therapies, individual or group, carried out with the children. In addition, there are Doctors' and Nurses' Groups whose fields of responsibility are self-evident.

Such a diversity of groups constitutes the "machinery" of a community. Needless to say, the oil that makes the "machinery" work is inter-communication and mutual interest. Where these slacken, the machinery grinds.

Lastly, there is the question of living-together as it pertains to the staff, which is the most difficult question upon which hinges the rise and fall of any community.

One can frequently observe that, whereas a group of adults will go out of its way to extend compassion and empathy to children, it will be wellnigh disinclined or unable to do the same for itself. This is, no doubt, as it should be, because the group must assume that its members are fit and not in need of the therapeutic understanding it wishes to extend to those committed into its care as children. Justifiably, there is the healthy attitude: We're not here for ourselves, but for the children; let's get on with the job.

But in an ever-increasing therapeutic, tolerant and permissive age, getting-on-with-the-job is not so easy. The move from religion to psychology, in Jung's terms, has had its effect. Self-denial and self-castigation for the sake of an ideal for which St. Francis of Assissi was so beautiful an example, is no longer really possible today. Our *persons* have been released from their confined underworld by the forces of pscho-analysis and psychology, and these *persons* state their needs to be comforted, to be understood, and to be loved, if they are in their turn to comfort, understand, and love others. Self-denial has given way to self-search.

Therefore, much greater skill and therapeutic determination are required for the art of adult living-together than for living with handicapped children who appeal so immediately to our compassion and therapeutic enthusiasm.

One essential is not to fall into the trap of regarding and reacting to another person in terms, as it were, of his constitution, for this creates a vicious circle. Not only does a person's physical as well as emotional constitution inflict itself on others; it inflicts itself to a far more existential degree on the person himself. This is indeed our lasting personal struggle—the dichotomy and contention between what we feel we *really* are or could be, and the burdensome features of our physical, emotional and to some extent our mental constitution.

When one physical-emotional constitution regards and reacts to another physical-emotional constitution, both are caught in the wheel of given circumstances, and both lose sight of the essential means of release, which is the other person's *potential* being, something that in rare moments can be perceived and experienced by others, and hardly ever of the Self by the Self. When the Holy Spirit descended on the heads of the disciples at Pentecost, none perceived the tiny tongue of fire that hovered over his own head, but each saw it hovering over the heads of the others.

This is the ultimate iron rule for community-living—the constant reference to the other person's spiritual potential, even if it remains hidden, for the question must be put: Do I not see the other person's spiritual potential because he has none or because he hides it? Or: am

I perhaps inhibited by some blindness of my own that does not allow me to confess to the other's potential?

Techniques and methods of living-together are being defined and put into practice in therapeutic and other communities today, but technique and method alone do not lead beyond a certain horizon in human existence.

Basically, there is a profound fear of loving the spiritual potential of our fellowmen, for if we would do so unreservedly, we would re-enter Paradise, and life on this stricken, precious, terrible and beloved earth would lose its purpose. Our mutual antipathies, irritations, misunderstandings, inquisitions are our plea to the Ultimate Reality to allow us a little more time.

We need not fear perfection in our relation to handicapped children. We are glad to be able to rectify even in small measure what has been so impaired. Human affliction makes us realists. But human potential is disturbing. We long for nothing more, but at the same time, we fear it and in so doing, deny it in the other, destroy it in the other, and cling tenaciously to the other's constitutional behaviour.

Once an individual has committed himself to community-living, he cannot disclaim responsibility for the others in the community, just as little as he can deny to the others a feeling of responsibility toward himself. There must be a constant sorting-out between the purely individual, and the individual inasmuch as he is a member of the group. No community can be instituted once and for all; it has to be created anew each day. The building-stones are the individual efforts to realise and make effective the ideals common to all the community.

Now as to the third sphere of community life—that of the "Body" to which I related work and its economic implications, I should like to describe the experiment carried out by this community since its inception thirty years ago, knowing that the economic side of community-living is an essential as well as fascinating challenge which will only come into its own when the world has advanced to a recognition of what human, social and spiritual elements are at work in Economics.

Meanwhile, the experiment of this small community is as follows. No salaries are taken by its members in the attempt to realise a principle according to which the welfare of a community is greater to the extent that an individual does not claim the rewards of his labour for himself, but when his needs are met by the fruits of the labour of the others.

Individual needs vary with individuals and at different times of life. For example, a young couple with a growing family and who may in no way be as yet expert, will require more to maintain itself than the senior expert whose family has grown up and has become independent. A senior staff member will find that his personal requirements tend to become reduced the older he grows, but he will be in demand as a spokesman or lecturer and therefore, his travelling expenses on behalf of others will increase, whereas the personal requirements of a younger member will be greater and his "public" expenses far less. In addition, the requirements of individuals in themselves are extremely variable in accordance with temperament, inclination, way-of-life, sense of values and so on. Naturally, when salary-scales have to be worked out, these things cannot be taken into consideration. In a community where no salaries are taken, they assume a different significance. Not that each individual can have what he or she likes. The individual must at all times distinguish between *wish* and *need* and in any case, must move within the confines of what the overall financial position of the community allows.

Incoming monies which, in this case, are the fees payable by local authorities in respect of each child sent to the school, are paid into a central account and from there, an allocation drawn up on the basis of budgeting is made to each of the house-communities, which in turn deals with the needs of the individuals who make up the staff, with the necessary running expenditure of children and adults in the house, maintenance, staff holidays and everything else pertaining to that particular group.

A certain percentage of the fee income is retained for overheads such as rates, etc., and for development. The community has no grants and no financial backing other than school fees and sums raised through the efforts of parents. The building programme, which is considerable, has been carried out largely from the fee income.

The community is officially recognised as a charity not being in the position to pay income tax and national insurance as its members have no personal income. For this reason, the permanent members of the community forgo old-age benefits. There is, therefore, no age for retiring and those who are advanced in years or frail in health will contribute to the work in their own way and will be maintained like everyone else in the community.

The argument is frequently brought forward that the individual lives in a false paradise, removed from the realities of the world outside, where people must contend with rates, rents, insurances and the like. The answer to this argument lies in the participation and responsibility of each member of the community, which means that he must extend to the monetary affairs of at least his own house-community, the concern he would have had to have for his own commitments and those of his dependents, had he a paid job outside. No doubt, there is the danger of apathy and non-involvement in a community, because the community will continue, so to speak, in spite of or without the participation of one or two individuals, whereas if they ceased to be responsible while holding paid jobs, they would lose their jobs. This is the challenge in community-building and community-maintaining, and we would not assume that an attempt of this kind can be expected to function of itself. It is hard work.

All accounting up to the moment when the books are presented to the auditor is done by the person in the particular unit in charge of the finances, who will discuss with his fellows in the house-community the financial situation. Single members have expense accounts to meet personal as well as household needs and each keeps a record of it.

Students may at times claim more extensive pocket money but with few exceptions, they prefer to fall in with the financial way of life of the community.

The close and intimate way of living with the children in such a community dissolves the question of working hours and shifts, and free time is arranged on an individual basis. Participation and responsibility are in principle unlimited, and can therefore be gauged in terms of wages only with great difficulty.

This does not mean that the members of this particular community are more self-sacrificing and more dedicated than others who work for handicapped children in different circumstances. It means that if all the consequences of community-living are taken, the result is bound to be some attempt to create a correspondence to community-living in economy. This particualr attempt did not stem from a simplified "Imitatio Christi", but rather as a social—and not altogether new—experiment in the conviction that the wage-system is a derivative of the theory of civilisations and that it will undergo a gradual evolution, signs of which can already be discerned, as there will be a shift in emphasis from a material standard of values to more essential human ones.

A community that wishes to uphold a differentiated and meaningful structure in which to carry out what it feels to be its mission must know how to replenish and rejuvenate its therapeutic, social and spiritual resources. It must know from where to take its new stimuli and how to recognise changing challenges. It must not withdraw from the world, yet it must be watchful lest its basic integrity be obscured, or watered-down. This requires study, heart-searching, realism and endurance.

In conclusion, I should like to return to the children for and with whom, this community's work is done. In much, they have been the teachers of their teachers, the therapists of their therapists, and have led the way in this attempt at new community-living. Because it is not only the handicapped child who derives his maturity from the experience of being able to help others, it is also we who need to be allowed to help others in a meaningful way if we are to come to the fulfilment of our own mature existence. A school for developmentally handicapped children should therefore be based on this principle of mutual help, on the principle of complete non-segregation, be it in the realm of age groups, in the realm of different handicaps, or between children and staff. It must be a living together, a giving and taking, in which the teacher consciously realises that he can only be a teacher because there are children who can still learn, in which the physician realises that he can only be a doctor because others are ill, in which the adult knows his own integrity and maturity comes to him from the relative development and success of the handicapped child entrusted to his care.

3. Adult Communities

Adult life for a handicapped person remains today a serious and unsolved question, which looms ominously from early infancy. Even if parents find a tolerant school which will take their handicapped child through his growing years, there is no guarantee that he will cope or be coped with when he is adult other than in the total protection provided for the sub-normal.

The anxiety that results gives rise to the feeling in parents and doctors that it is better to place a handicapped child in secure protection from which he will not have to be discharged when he is adult, rather than press him through schooling, only to have to face the question, "What now'?"

Anxiety and insecurity are characteristic of our time and because of this, there is an intense urge to make "lasting provisions" or find "final solutions" for handicapped people. Yet life itself defeats this call for final solutions or lasting provisions, because life consists of change and development, and has no final solutions, with the possible exception of death.

The very wish to make lasting provisions for the mentally handicapped has made mental sub-normality hospitals into places of living death. However, the understanding of and the attitude to the problem of the adult handicapped person have begun to change, and with this change, new possibilities have opened up.

Undoubtedly, of the more severely handicapped and disturbed children who manage to attend remedial schools, only a relatively small number will cope independently with adult life. The great majority will require the support of a tolerant environment which is geared to the specific needs of the adult handicapped person. Here, the trend in therapeutic communities has a considerable contribution to make, a point to which I shall return later.

Among handicapped school-leavers, there will be a number up to 20 per cent who will not even manage in the sheltered facilities provided in the wider community, but will require that protection which only a family, or failing this, a mental sub-normality hospital can provide. But here, it is important to realise that neither of these is or needs to be a permanent or final state.

A handicapped person may work for a number of years in a

sheltered therapeutic community, and then at some point in his life, decide to leave it and find employment on the open market. This decision can be determined by a number of factors. The country as a whole may be going through a period of shortage of labour so that the chances of employment for the handicapped are more favourable. Another factor is that a handicapped person, having established himself, no longer requires the degree of tolerance and shelter he previously needed, and wishes to fend for himself. Equally, the therapeutic community itself might be undergoing a change and see the necessity of catering for the needs of more severely handicapped and disturbed persons, so that the more able person feels out of place.

On the other hand, a handicapped person may have managed in ordinary employment and lived on his own and yet, have reached a stage when he requires, for a period, the greater security offered by a therapeutic community. Or, a person who has been an inmate in a mental sub-normality hospital may later become sufficiently rehabilitated to spend a further period of his life in a therapeutic community. The reverse, too, may and does happen. Even in the life of one individual, there may have to be an alternating between the therapeutic community and the mental sub-normality hospital.

I believe that it is of paramount importance to maintain an *open* attitude to the situations of handicapped adults. We do the greatest, and possibly the most unforgivable injury to a handicapped person when we consider his case closed. This is tantamount to murder.

There has been a considerable break-through in the field of psychiatry and an increasing trend in hospitals for the mentally ill to rehabilitate their patients for the life in the wider community as soon as possible. On a much smaller scale, the same trend has been discernible in the field of mental handicap throughout the last two decades.

I shall now describe a therapeutic community for mentally handicapped adults which has been working for a number of years. When the idea of this community was conceived, the fact that there has been and is a general move of the population from the country to urban areas played a determining part in the realisation of the idea; the de-populated areas with small upland vacated farms seemed both to attract and to require some such venture as that of a therapeutic community. Needless to say, in choosing an increasingly depopulated rural upland area, agriculture has played a predominant part in the community from the beginning.

This particular community, however, was not conceived as a therapeutic community or as a kind of "open" sub-normality unit in disguise. It was conceived as a rural village with all the ramifications that pertain to village life, and although this is still developing, I shall call it a village-community instead of a therapeutic community, as this underlines the basic intention. I should like to add that this is not the only village-community of the sort, but I wish to describe this one as a "prototype".

To begin with, something should be said about "admission" to the village community. Parents frequently enter their children's names on the waiting list long before they have reached adulthood. This does not mean that the village-community is under an obligation to take a person whose name has been on the waiting-list for so long, because the community recognises the wish of the young adult concerned to live and work in such a community as a decisive factor in admission. In the absence of his own choice, his status is by necessity that of a patient.

The fact that a parent enters his child's name on the waiting-list of the village-community stems from the understandable wish of the parent to secure the future of his child. But the way a child grows up into adulthood may take a very different course from that which a parent may envisage and hope for. Upon leaving school, an adolescent handicapped person may be well-equipped to hold a relatively responsible job, but his particular environment may have nothing to offer, or his parents' anxiety as to his future may so frustrate his efforts to adapt to "normal" society that there is a total breakdown of continuity and the person himself, although originally capable, may have to be admitted to a mental hospital.

When he is then referred to the village-community for admission, he will have had a fair experience of frustration and disappointment which will enable him to make his own considered opinion whether to try life in the community, or not.

Equally, handicapped school-leavers often have very illusory ideas, sometimes supported by their parents, of their own capabilities. It is essential that they too, face the disillusionments and disappointments that lie ahead, because only by so doing, can they arrive at a more realistic assessment of their own disabilities as well as the disabilities of their home-environment in the failure to integrate them in a workable way. It is important to realise that illusions are not only the prerogative of the "normal", but a formidable and obstructive factor in the realistic integration of handicapped persons.

Even the non-speaking, autistic young person whom no one thinks capable of independence, needs to face the impasse his environment experiences in seeing him as an adult person with his peculiarities in its midst.

All this is necessary experience, both for a handicapped person as well as for his family, and enables the person to acquire a maturity of personal decision as to whether he wants to enter a village-community or not. Therefore, these village-communities are *not* continuations of provisions offered to the mentally handicapped in childhood.

The decision to join a village-community on the part of a handicapped adult implies that he is willing to be socially co-operative. A very able and skilful person who has a degree of independence may fail entirely in a village-community because of his anti-social and negativistic outlook. On the other hand, a person with a considerable physical handicap which reduces his working contribution may be a very successful member of the community owing to his sense of social responsibility. Again, in an individual, social ability may fluctuate. He may go through a phase of severe withdrawal, in which case, he may have to leave for a period, if the community does not see its way to contain him until the phase is over.

This is to illustrate the fact that the village-community, although relatively non-specific as far as types and degrees of handicap and disturbance are concerned, is in no way a *final* solution.

No fees are paid for people entering the community. Each handicapped person contributes what he can by way of work to the maintenance of the entire community which, in turn, meets his needs.

What makes it difficult for a handicapped person to live in ordinary competitive society is not so much his relative inability to work skilfully, but rather his inability to handle his wages and personal commitments. For this reason, the Ministry of Employment and Productivity found it exceedingly difficult to include the mentally handicapped in its otherwise very effective Disablement Scheme.

One solution to this problem can be found in the village-community in which everyone, including the staff, lives on a non-salary basis. Income is shared according to need in the various households and a social coherence established which transcends any distinction between staff and handicapped. This allows for an administrative structure in which all members of the community participate on the basis of their abilities and social interests. Hence, a handicapped person does not have to deal with his own isolated pay-packet (many handicapped people attach little value to pay-packets and tend to "donate" them to others), but is part of a wider and vital socio-economic complex in which he plays an important role.

Under these circumstances, the Ministry of Employment and Productivity recognises the village-community as a sheltered workshop under its Disablement Scheme and pays a Deficiency Grant for the majority of handicapped adults. Those for whom the Ministry does not accept responsibility are maintained by the community.

There is a considerable amount of charitable activity on behalf of the village-community and the proceeds of this are no longer used to meet running expenditure, but development and building.

Production is the other source of income which, together with the Deficiency Grants, meets the running expenditure of the village-community. Here, the farms have reached a good and reliable level of production which includes dairy-farming and pig-breeding, and besides, there is a considerable variety of small industries such as baking, weaving, woodwork, pottery, soft-toymaking, glass-engraving, jewellery, enamel work, batik, basketry, candle-making and the like.

Where applicable, division of labour is used in the workshops to create production-processes simple enough to be carried out efficiently by the more severely handicapped, as well as for reasons of productivity as such.

A high standard of quality has been striven for and is maintained. Articles produced by the village-community are in demand in the wider community. Firms and stores in larger cities place large standing orders which have to be met constantly. Hence, there is an intense working atmosphere in the community, with definite working hours and working discipline.

In the single workshops, the workers participate in the production of articles from the state of raw material to the finished article ready for the Sales Department. They know by whom their products are required and where they go. To work to satisfy the requirements of others, together with achieving a sense of fulfilment that comes with making a beautiful and useful article, is the primary motivation in each workshop group. The motivation of working for self-support or gain hardly enters the picture.

Fundamentally, a man does not work in order to earn his living. He works because he *needs* to work in order to achieve self-expression and self-fulfilment. But our society today does not recognise man's fundamental relation to work as self-fulfilment. Its professed motivation is gain and affluence. In this, a handicapped person has no viable place. He must experience the existential meaning of work, if he is at all to establish himself as a person.

These are issues which are discussed in the workshop groups, and it is astonishing how the productive capacities of handicapped and disturbed people develop as a result.

In parenthesis, it should be said that the often remarkable skills and gifts of autistic and psychotic persons undergo a considerable degree of "normalisation" or "rationalisation" in village-workshop life in that these skills move from being fixations and obsessions into being assets in production. There is evidence that this is not merely a construed justification of the use of skills in psychotic persons, but that such a person himself experiences release when his obsessional compulsions become useful in productivity. It is an interesting fact that a person who was severely withdrawn or psychotic as a child frequently represents a particular social element in an adult working community. He is the upholder of punctiliousness, conscientiousness and exact workmanship and is often the one who, along with the craft master, holds the workshop together.

The social side as distinct from the socio-economic one is almost the more essential side of village-community life. In a community for handicapped children, there is a natural distinction between children and adults, the latter being teachers and therapists, the former being pupils and patients. These distinctions fall away in the village-community and with them, the last vestiges of pedagogical and remedial aspiration. In other words, handicapped adults in the village are no longer "educated" or "treated". For example, there is a deliberate abandonment of "case conferences" which play such a central part in work with children. An adult must be accepted as he is and his dignity as a person upheld. He knows that he is handicapped, because he helps to run the village-community the purpose of which is manifest, and he will accept his situation with surprising maturity and candour.

As I have said, no distinction is made between staff and "patient"; all are adults some more handicapped and peculiar than others, living and working in a generally therapeutically orientated community.

This emphasis on non-distinction means that the staff must share *everything* with the handicapped adults with whom they live, a thing which requires tact and realism as it can neither be enforced nor simulated for the sake of a fine idea. The equal economic status of everyone in the village-community is a sound and workable basis for sharing life in other spheres of a more individual and personal nature, for example, holidays.

The many administrative groups such as Production, Finance, Land, Cultural Activities and the like contain both staff and handicapped, not nominally, but on the strength of useful contribution. Some vital activities and "public services" in the village-community are sustained purely by handicapped people, a number of whom are expert at things necessary to the life of the community, while others are not.

When it comes to the question of marriage between two handicapped people, the principle of non-distinction between staff and handicapped comes up against a specific test. Some marriages have taken place and the children of these marriages grow up along with the staff children and go to school in the little "village school" which has been established for children in the community. Marriages are neither encouraged nor discouraged. Each time the issue arises, it is gone into carefully and candidly with those concerned and with as little illusion as possible as to what is at stake.

Cultural activities in the village-community are manifold. Lectures, talks, study groups, drama groups, choir and folk-dancing are on the weekly programme of the Community Centre. These activities are not "laid on", but are vital expressions of community living. And finally, spiritual cohesion is centred on the Chapel as the spiritual focal point in the intense working and cultural life of the community.

Village-communities like the prototype I have described are attracting large numbers of people suffering from the most diverse developmental handicaps, but also highly intelligent people who, because of neurosis or mental illness, seek shelter and tolerant companionship. These village-communities have in addition, begun to attract young people in search of new ways of community living and social service. They are attracting older people who, having retired after a full life, settle in the villages and place their experience at the disposal of the community.

The village-communities for the mentally handicapped have begun to establish contact with the therapeutic communities in psychiatric hospitals to the benefit of both communities. Just as they have begun to extend help in the field of psychiatry, so also the time is coming when they will include the very severely disabled who meantime require care in mental sub-normality hospitals. Many handicapped persons make ideal nurses and are the right persons to help those more afflicted than themselves.

As tolerance towards human frailty grows today, a time in the not too distant future can be envisaged when such therapeutic communities or village-communities will offer the remedy for some of the social, spiritual and economic ills of the present-day. Certain prominent men have suggested that communities of those who cannot keep up with the competitive life of today may well be forming in seclusion, as eddies in the broad stream of human civilisation and are seeds for a future culture.

If I may be permitted to touch briefly upon an ideal which is at the same time a cherished urge, I should like to call attention to the many forsaken upland areas in the country and to lonely glens which seem to be waiting for a resurgence of life and cultivation. If village-communities for the mentally handicapped and therapeutic communities for the psychiatrically ill would take hands with Forestry Experts and Land Conservationists and others, perhaps some of these virtual wastes in the country might flourish again.

We must be aware of the fact that developmental handicap is not merely the result of illness and misfortune. It is also the outcome of the advancements and triumphs in the field of medicine and science, as well as of the increasingly tolerant attitude of people today. Handicaps and developmental failures have always existed, but in the past, afflicted children rarely survived birth. Those who did survive were put away out of sight.

Scientific and social advancement has drawn the handicapped out into the light of day. Like Lazarus who rose out of the grave, handicapped children have stepped forward into our midst. Perhaps they too can lead us to new life.

Bibliography

PART I

CLARKE, A. D. B., and CLARKE, A. M. *Mental Deficiency in the Changing Outlook.* London: Methuen, 1958.

CLARKE, A. D. B. *Recent Advances in the Study of Subnormality.* Nat. Assoc. for Mental Health, 1966.

CRAFT, M., and MILES, L. *Patterns of Care for the Subnormal.* Pergamon, 1967.

DEGENAAR, A. G. (Ed.). *Zur Heilpaedagogik.* Basel. Zbinden and Hugin, 1938.

FABER, N. W. *The Retarded Child.* New York: Crown, 1968.

FURNEAUX, B. *The Special Child.* Penguin Ed. Spec., 1969.

GEUTER, I. *Adventure in Curative Education.* New Knowledge Books, 1962.

ST. CHRISTOPHER'S SCHOOL. *In Need of Special Care.* Bristol, 1967.

KONIG, K. "The Care and Education of Handicapped Children", in: *Aspects of Curative Education* (Ed. Pietzner). Aberdeen University Press, 1966.

KORCZAK, J. *Wie man ein Kind lieben soll.* Gottingen: Vandenhoeck and Rupprecht, 1967.

LEWIS, M. M. *Language, Thought and Personality in Infancy and Childhood.* Harrap, 1963.

MASLAND, R. L., SARASON, S. B., and GLADWIN, T. *Mental Subnormality.* New York: Basic Books, 1958.

MINISTRY OF EDUCATION. *The Health of the Schoolchild.* H.M.S.O., 1954.

PACHE, W. (Ed.). *Heilende Erziehung.* Natur Verlag Arlesheim (Schweiz), 1956.

PIETZNER, C. (Ed.). *Aspects of Curative Education.* Aberdeen University Press, 1966.

RUDEL, J. "The Challenge of the Handicapped Child", in: *The Faithful Thinker.* Hodder and Stoughton, 1961.

SEGAL, S. S. *No Child is Ineducable.* Pergamon, 1967.

STEINER, R. *Heilpaedagogischer Kursus 1924.* R. Steiner Nachlass Verwaltung. Dornach, Schweiz, 1965.

STEVENS, and HEBER. *Mental Retardation.* University of Chicago, 1964.

PART II

ASPERGER, H. *Heilpaedagogik.* Wien: Springer, 1956.

BOWLBY, J. *Maternal Care and Mental Health.* W.H.O. Monograph No. 2.

CARMICHAEL, L. (Ed.). *Manual of Child Psychology.* Chapman and Hall, 1954.

ERIKSON, E. H. *Childhood and Society.* London: Imago, 1951.

FALKNER, F. (Ed.). *Human Development.* Saunders, 1966.

FREUD, A. *Normality and Pathology in Childhood.* Hogarth Press, 1965.

FROMMER, E. A. *Voyage Through Childhood into the Adult World.* Pergamon, 1969.

GARRISON, KINGSTON, BERNARD. *The Psychology of Childhood.* Staple Press, 1967.

GESELL, A. (Ed.). *The First Five Years of Life: A guide to the study of the pre-school child.* Methuen, 1954.

GESELL, A., and ILG, F. G. *The Child from Five to Ten.* Hamish Hamilton, 1946.

HOWELLS, J. G. (Ed.). *Modern Perspectives in Child Psychiatry.* Oliver & Boyd, 1965.

ILLINGWORTH, R. S. *The Development of the Infant and Young Child; normal and subnormal.* Livingstone, 1966.

JACOBI, J. *The Psychology of C. G. Jung.* Routledge and Kegan Paul, 1968.

KAHN, J. H. *Human Growth and the Development of Personality.* Pergamon Press, 1965.

KONIG, K. *The First Three Years of the Child.* New York: Anthroposophic Press, 1969.

KÖNIG, K. *Brothers and Sisters, a Study in Child Psychology.* Mass: St. George Book Service, 1963.

LUTZ, J. *Kinderpsychiatric.* Zurich: Rotapfel, 1964.

MERLEAU PONTY. *Primacy of Perception.* Edie (Ed.) North-west University Press, 1964.

MERLEAU PONTY. *Phenomenology of Perception.* Routledge and Kegan Paul, 1962.

MOOR, P. *Heilpadagogische Psychologie.* Bern: H. Huber, 1951.

NITSCHKE, A. *Das Verwaiste Kind des Natur.* Tubingen: Niemeyer Verlag, 1962.

PIAGET, I., and INHELDER, B. *The Psychology of the Child.* Routledge and Kegan Paul, 1969.

SODDY, K. *Clinical Child Psychiatry.* Baillière, Tindall and Cox, 1960.

SPITZ, R. A. *The First Year of Life. A Psychoanalytical Study of Normal and Deviant Development of Object Relations.* New York: International University Press, 1965.

STEINER, R. *The Education of the Child from the Anthroposophical Point of View.* Rudolf Steiner Publishing Co., 1947.

STEINER, R. *Study of Man.* Rudolf Steiner Press, 1966.

STEINER, R. *The Kingdom of Childhood.* Rudolf Steiner Press, 1964.

STERN, W. *Psychologie der Fruehen Kindheit.* Leipzig, 1928.

TANNER, J. M., and INHELDER, B. *Discussions of Child Development.* Geneva 1953/4/5/6. Tavistock Publ., 1963.

WEGMAN, I. *Principles of Curative Education.* Anthroposophical Publishing Co., 1942.

WERNER, and KAPLAN. *Symbol Formation*. New York: J. Wiley & Sons, 1963.
WINNICOT, D. W. *The Family and Individual Development*. Tavistock Publ.,
1965.
WOLFF, S. *Children under Stress*. Allen Lane, 1969.
ZULLIGER, H. *Bansteine Zur Kinderpsychotherapie*. Bern: H. Huber, 1966.

PART III

(i)

KÖNIG, K. "Some Fundamental Aspects of Diagnosis and Therapy in Curative
Education", in: *Aspects of Curative Education* (Ed. Pietzner). Aberdeen University Press, 1966.
MASLAND, R. L. "The Prevention of Mental Subnormality", in: Masland,
Sarason and Gladwin: *Mental Subnormality*. New York: Basic Books, 1958.
MAUTNER, H. *Mental Retardation, its Care, Treatment and Physiological Base*.
New York: Pergamon Press, 1959.
STEINER, R. *Study of Man*. London: Rudolf Steiner Press, 1966.
WEIHS, T. J. "Differential Diagnoses of Backward Children", in: *Aspects of
Curative Education* (Ed. Pietzner). Aberdeen University Press, 1966.

(ii)

DELCATO, C. H. *Treatment and Prevention of Reading Problems*. Illinois: Thomas,
1959.
HALLGREN, B. "Specific Dyslexia", *Acta Psychiatrica and Neurol.*, Supplement
65, Stockholm, 1950.
LUDWIG, W. *Das Rechts-Links-Problem im Tierreich und beim Menschen*. Berlin:
Springer, 1932.
MEER, VAN DER H. C. *Die Links-Rechts Polarisation des Phenomenalen Rammes*.
Groningen: Walters, 1959.
MONEY, J. (Ed.). *Reading Disability*. Baltimore: John Hopkins Press, 1962.
PEARCE, R. A. H. *Crossed Laterality. Archives of Disease in Childhood*. Vol. 28,
1953.
WEIHS, T. J. "Differential Diagnosis of Backward Children", in: *Aspects oj
Curative Education* (Ed. Pietzner). Aberdeen University Press, 1966.
ZANGWILL, O. L. *Current Status of Cerebral Dominance in Disorders of Communication*. (Eds. Rioch and Weinstein). Baltimore: Williams & Williams,
1964.

(iii)

ANDRE-THOMAS, and ANTGAERDEN, S. *Locomotion from Pre- to Post-Natal Life*.
Clinics in Developmental Medicine No. 24. Spastic Society and W. Heinemann Med. Books, 1966.

BENDA, C. E. *Developmental Disorders of Mentation and Cerebral Palsies*. New York: Grune and Stratton, 1952.

BOBATH, K., and BOBATH, B. "A Treatment of Cerebral Palsy", *British Journal Phys. Med.* 15.

DUBOVITZ, V. *The Floppy Infant*. Clinics in Developmental Medicine, No. 31. Heinemann Med. Books, 1969.

Hemiplegic Palsy in Children and Adults. Clinics in Developmental Medicine, No. 4. National Spastic Society, 1961.

HENDERSON, J. L. *Cerebral Palsy in Childhood and Adolescence*. Livingstone, 1961.

KONIG, K. "Some Aspects on the Treatment of Cerebral Palsy", *British Journal Physiotherapy*.

KÖNIG, K. "Dauerlaehmungen nach Poliomyelitis und Umwelt- schädigung", in: *Das Seelenpflege-bedürtige Kind*. II, 2, 1957.

SAHLMANN, L. "The Child Suffering from Cerebral Palsy", in: *Aspects of Curative Education* (Ed. Pietzner). Aberdeen University Press, 1966.

(iv)

ASPERGER, H. *Heilpaedagogik*. Wien: Springer, 1956.

BETTELHEIM, B. *Love is not Enough. The Treatment of Emotionally Disturbed Children*. Illinois: Free Press, 1950.

BIRCH, H. G. (Ed.). *Brain Damage in Children*. New York: Williams and Wilkins, 1964.

DAVENPORT, R. K. JR., and BERKSON, G. "Stereotyped Movements of Mental Defectives", *American Journal Mental Deficiency*, 67, 879, 1963.

ENGEL, KONIG, and MULLER-WIEDEMANN. *Ueber Schwere Kontaktstörungen im Kindesalter und deren Behandhung mit der Substanz Thalamos*. Stuttgart: Arbeitsgem. Anthrop. Aerzte, 1956.

ENGEL, P. *Mental and Associated other Sequela of Encephalitis in Infancy and Early Childhood*. n.p. 1957.

MULLER-WIEDEMANN, H. "The Postencephalitic Syndrome in Early Childhood", in: *Aspects of Curative Education* (Ed. Pietzner). Aberdeen University Press, 1966.

STRAUSS, and LETHINEN. *Psychopathology and Education of the Brain Injured Child*. New York: Grune and Stratton, 1951.

(v)

ANTHONY, J. "An Experimental Approach to the Psychopathology of Childhood Autism", *British Journal Med. Psych.*, 31, 1958.

BENDA, C. E. "Childhood Schizophrenia, Autism and Heller's Disease", in: Bowman and Mountners: *Mental Retardation*. New York: Grune and Stratton, 1960.

BENDER, L. "Autism in Children with Mental Deficiency", *American Journal Mental Deficiency*, 63, 1960.

BETTELHEIM, B. *The Empty Fortress.* Canada: Collier-MacMillan, 1967.

BOSCH, G. *Der Fruehkindliche Autismus.* Berlin: Springer, 1962.

CAMERON, K. "Psychosis in Infancy and Early Childhood", *Medical Press*, 234, 3, 1955.

CREAK, M. et al. "Schizophrenic Syndrome in Children", *British Medical Journal*, 2, 889, 1961.

CREAK, M. "Schizophrenic Syndrome in Childhood", *Developmental Medicine and Child Neurology*, 4, 530, 1964.

GOLDFARB, W. *Childhood Schizophrenia.* Harvard University Press, 1961.

HARMS, E. *Essentials of Abnormal Child Psychology.* New York: Julian Press, 1953.

KANNER, L. *Child Psychiatry.* Third Edition. Springfield: Thomas, 1957.

REED, G. F. "Elective Mutism in Children", *Journal Child Psychol. Psychiatry*, 4, 99, 1963.

RIMLAND. *Infantile Autism.* Methuen, 1965.

RUTTER, M. "The Influence of Organic and Emotional Factors on the Origins, Nature and Outcome of Child Psychosis", *Developmental Medicine and Child Neurology*, 7, 518, 1965.

SAHLMANN, L. "Autism or Aphasia", *Developmental Medicine and Child Neurology*, 11, 443–448, 1969.

STROH, G. *Psychosis in Childhood.* Public Health, 77, 21, 1962.

TAFT, L. T., and GOLDFARB, W. "Pre-natal and Perinatal Factors in Childhood Schizophrenia", *Developmental Medicine and Child Neurology*, 6, 32, 1964.

TRAMER, M. "Elektiver Mutismus bei Kindern", *Z. Kinderpsychiatric*, 1, 30, 1934.

WEIHS, T. J. "Psychotic Children", in: *Aspects of Curative Education* (Ed. Pietzner). Aberdeen University Press, 1966.

WING, J. K. (Ed.). *Early Childhood Autism.* Pergamon Press, 1966.

(vi)

BLANK, H. R., and ROTHMAN, R. "The Congenitally Blind Child", in: *Proceedings Inst. Soc. Serv. Org.* New York: Guild Jewish Blind, 1952.

FARREL, G. *The Story of Blindness.* Cambridge: Harvard University Press, 1956.

GARDINER, MACKEITH, and SMITH. *Aspects of Developmental and Paediatric Ophthalmology.* Spastics Med. Publ., 1969.

"Guidance and Treatment of Blind Children", in: *Aspects of Curative Education* (Ed. Pietzner), pp. 240–249. Aberdeen University Press, 1966.

LOEVENFELD, B. *Our Blind Children.* C. C. Thomas, 1956.

LOEVENFELD, B. (Ed.). *The Blind Pre-School Child.* New York: American Foundation for the Blind, 1947.

NORRIS, SPALDING, and BRODIE. *Blindness in Children.* University of Chicago, 1957.

PRINGLE, M. L. K. *The Educational and Social Adjustments of Blind Children.* N.F.E.R. Publ. No. 10, 1964.

SCHUMANN, H. *Träume der Blinden.* Basel: S. Karger, 1959.

STEINBERG, W. *Vom Innenleben der Blinden.* München: Reinhardt, 1955.

(vii)

ARNOLD, T. *Education of Deaf-Mutes.* London: Wertheimer, 1888.

BROWD, V. *New Way to Better Hearing.* Faber and Faber, 1953.

EWING, A. and E. *Teaching Deaf Children to Talk.* Manchester: University Press, 1964.

EWING, I. and A. *Opportunity and the Deaf Child.* University London Press, 1947.

HODGSON, K. W. *The Deaf and Their Problems.* Watts & Co., 1953.

KERN, E. *Ganzheitlicher Sprachunterricht für das Gehörgeschaedigte Kind.* Freiburg: Hender, 1958.

KONIG, K. "Music Therapy in Curative Education", in: *Aspects of Curative Education* (Ed. Pietzner). Aberdeen University Press, 1966.

MULLER-WIEDEMANN, S. "Aus der Arbeit mit Tauben Kindern", in: *Das Seelenpflege bedürtige Kind.* (Ed. Starke). I. 2, 1955.

MYKLEBUST, H. R. *Auditory Disorders in Children.* New York: Grune and Stratton, 1954.

WHITNALL, E., and FRY, D. B. *The Deaf Child.* Heinemann Med. Books, 1964.

(viii)

BRAIN, LORD. *Speech Disorders.* London: Butterworth, 1965.

CRITCHLEY, M. *Developmental Dyslexia.* Heinemann, 1964.

CROSBY, R. M. N. *Reading and the Dyslexic Child.* Souvenir Press, 1968.

FRANKEIN, A. W. *Children with Communication Problems.* Pitman, 1965.

HEAD, H. *Aphasia and Kindred Disorders of Speech.* Cambridge University Press, 1926.

HOCH, P. H., and ZUBIN, J. (Ed.). *Psychopathology of Communication.* New York: Grune and Stratton, 1958.

KONIG, K. *The First Three Years of the Child.* New York: Anthroposophic Press, 1969.

MASON, S. (Ed.). *Signs, Signals, Symbols.* Methuen, 1963.

MORLEY, M. S. *Development and Disorders of Speech in Childhood.* Livingstone, 1965.

ORTON, S. T. *Reading, Writing and Speech Problems in Children.* New York: Norton & Co., 1937.

PARREL, S. *Speech Disorders.* Pergamon, 1965.

REUCK, and O'CONNOR. *Disorders of Language*. (Ciba Foundation Symp.) London: Churchill Ltd., 1964.

RIOCH, D., and WEINSTEIN, E. (Ed.). *Disorders of Communication*. Baltimore: Williams and Williams, 1964.

SAHLMANN, L. "Autism or Aphasia", *Developmental Medicine and Child Neurology*, 11, 443–448, 1969.

STEINER, R. *Anthroposophy, Psychosophy, Pneumatosophy*. 1909. Dornach: Philos. Anthrop. Verlag, 1931.

(ix)

ANDRY, R. *Delinquency and Parental Pathology*. Methuen, 1960.

BETTELHEIM, B. *Love is not Enough. The Treatment of Emotionally Disturbed Children*. Illinois: Free Press, 1950.

BETTELHEIM. B. *Truants from Life. The Rehabilitation of Emotionally Disturbed Children*. Illinois: Free Press, 1955.

BOWLBY, J. *Child Care and the Growth of Love*. Pelican Books, 1951.

BRIDGE, E. M. *Epilepsy and Convulsive Disorders in Children*. New York: McGraw & Hill, 1949.

CAPLAN, G. *Prevention of Mental Disorders in Children*. Tavistock Publ., 1961.

CRAFT, M. J. "Delinquency, Mental Disorder and Dullness", *British J. Criminal*, 5, 55, 1962.

FERGUSON, T. *The Young Delinquent in his Social Setting*. Oxford University Press, 1952.

GLOVER, E. *The Roots of Crime*. London: Imago, 1960.

HARMS, E. *Essentials of Abnormal Child Psychiatry*. New York: Julian Press, 1953.

HARMS, E. (Ed.). *School-Psychopathology and Classroom-Psychotherapy. The Nervous Child*, Vol. 10, No. 3–4. New York: Child Care Publ., 1954.

HEYMANN, K. *Infantilismus*. Psych. Praxis 16. Basel: Karger, 1955.

HOCH, P. H., and KNIGHT, R. P. (Eds.). *Epilepsy. Psychiatric Aspects of Convulsive Disorders*. Heinemann, 1948.

HOME OFFICE. *The Child, The Family and the Young Offender*. H.M.S.O., 1965.

JANTZ, D. *Die Epilepsien*. Stuttgart: Thieme Verlag, 1969.

LEWIS, M. *Deprived Children*. Oxford University Press, 1954.

MAKARENKO, A. *The Road to Life*. Moscow: Foreign Publ. House, 1955.

MINISTRY OF EDUCATION. *Report of the Committee on Maladjusted Children*. H.M.S.O., 1955.

STEINER, R. *Heilpaedagogischer Kurs*. 1924. Dornach: R. Steiner Nachlassverein, 1965.

STOTT, D. H. *Studies of Troublesome Children*. Tavistock Publ., 1966.

WINNICOT, D. W. *The Child, The Family and The Outside World*. Tavistock, 1964.

WOLFF, S. *Children under Stress.* Allen Lane, 1969.
WOODWARD, M. *Low Intelligence and Delinquency.* London: Inst. for Sci. Treatment of Delinquency, 1955.

(x)

BURT, C. *The Causes and Treatment of Backwardness.* University London Press, 1953.
BURT, C. *The Subnormal Mind.* Oxford University Press, 1955.
CLARKE, A. D. B. *Recent Advances in the Study of Subnormality.* Nat. Assoc. for Mental Health, 1966.
EDMUNDS, L. F. *Rudolf Steiner Education.* Rudolf Steiner Press, 1962.
GALLAGHER, J. J. *The Tutoring of Brain Injured Mentally Retarded Children.* Illinois: C. C. Thomas, 1960.
HARWOOD, A. C. *Recovery of Man in Childhood.* Hodder and Stoughton, 1958.
HEYMANN, K. *Heilpaedagogisches Lernen.* Psychol. Praxis 26. Basel: Karger, 1960.
KIRCHOFF and PIETROWICZ (Eds.). *Konzentrations-schwache Kinder.* Psychol. Praxis, 24. Basel: Karger, 1959.
KLOSS, H. *Waldorf-Paedogik und Staatschulwesen.* Stuttgart: Klett Verlag, 1955.
LEYS, D. *The Needs of Mentally Handicappᵣd Children.* London: S. E. Metropolitan Hospital Board, 1962.
LUNIE, A. *The Mentally Retarded Child.* Pergamon Press, 1963.
PRITCHARD, D. G. *Education and the Mentally Handicapped.* Routledge and Kegan Paul, 1963.
PIETZNER, C. (Ed.). "Education of Handicapped and Disturbed Children. A Report", in: *Aspects of Curative Education.* Aberdeen University Press, 1966.
STEINER, R. *The Kingdom of Childhood.* Rudolf Steiner Press, 1964.
STEINER, R. *The Essentials of Education.* Rudolf Steiner Press, 1968.
STEINER, R. *Education and Modern Spiritual Life.* London: Anthrop. Publ. Co., 1954.
STEVEN, M. *Observing Children who are Severely Subnormal.* E. Arnold, 1968.
WESTON, P. T. B. *Some Approaches to Teaching Autistic Children.* Pergamon Press, 1965.

(xi)

BENDA, C. E. *The Child with Mongolism.* New York: Grune and Stratton, 1960.
CARTER, C. O. *et. al.* "Chromosome Translocation as a Cause of Familial Mongolism", *Lancet,* Vol. 2, 678–680, 1960.
INGELLS, T. H. "Etiology of Mongolism", *American Journal Dis. Children,* Vol. 74, 147–165, 1947.
KONIG, K. *Der Mongolismus.* Stuttgart: Hyppokrates Verlag, 1959.

MAUTNER, H. *Mental Retardation: Its Care Treatment and Physiological Base.* New York: Pergamon, 1960.

PENROSE, L. S. "Maternal Age in Familial Mongolism", *Journal Mental Science*, Vol. 97, 738–747, 1951.

PENROSE, L. S. "Observation on the Aetiology in Mongolism", *Lancet*, Vol. I, pp. 505–509, 1954.

PENROSE, L. S. "Chromosomal Translocation in Mongolism and in Normal Defectives", *Lancet*, Vol. II, pp. 409–410, 1960.

WOLSTENHOLME and PORTER (Eds.). *Mongolism.* Ciba Foundation No. 25. J. & A. Churchill, 1967.

PART IV

BOWLBY, J. *Maternal Care and Mental Health.* W.H.O. Monograph Series No. 2.

ERIKSON, E. H. *Childhood and Society.* Imago, 1951.

FERGUSON, T., and KERR, A. W. *Handicapped Youth.* Oxford University Press, 1961.

GARTNER, M. "Mental Handicap", in: *Aspects of Curative Education* (Ed. Pietzner). Aberdeen University Press, 1966.

O'CONNOR, N., and TIZARD, J. *The Social Problem of Mental Deficiency.* Pergamon Press, 1956.

STEINER, R. *Geistegwissenschaft und Soziale Frage.* Dornach: R. Steiner Nachlass-Verwaltung, 1957.

TIZARD, J., and GRAD, J. C. *The Mentally Handicapped and their Families.* Oxford University Press, 1961.

WEIHS, T. J. "Family of the Handicapped", in: *Mental Health*, Spring, 1969.

WEIHS, T. J. *Superintendent's Report.* The Camphill–Rudolf Steiner–Schools, Aberdeen, Scotland, 1955–62.